American Institute of Chemical Engineers

Topical Conference on Energy Processes 2006

Held at the 2006 AIChE Spring National Meeting

April 23-27, 2006
Orlando, Florida, USA

Printed from e-media with permission by:

Curran Associates, Inc.
57 Morehouse Lane
Red Hook, NY 12571
www.proceedings.com

ISBN: 978-1-60423-529-6

Some format issues inherent in the e-media version may also appear in this print version.

American Institute of Chemical Engineers

Topical Conference on Energy Processing
2006

TABLE OF CONTENTS

Session 64 - Simulation of Heat and Mass Transfer Equipment in Process Plants

Chair: Vishwas V. Wadekar

Session 95 - Low Grade Heat Utilization

CoChair: C.B. Panchal

Session 96 - Microfluidic Devices for Energy Applications Tutorial

CoChair: Yangchuan Xing

Chair: Kim Henthorn

115 **Modeling and Evaluation of an Integrated Microfluidic Fuel Processor for Miniature Power Sources**
Dongying Qian, R.S. Besser, Adeniyi Lawal

127 **Air-Breathing Laminar Flow Fuel Cells Operating in Alkaline or Acidic Media**
Paul J.A. Kenis, Ranga S. Jayashree, Daniela Egas, Dilip Natarajan, Jacob Spendelow, Larry J. Markoski

128 **Addressing Critical Issues in Microscale Fuel Processing: Demonstration of an Integrated Silicon Microreactor Based Methanol Steam Reformer**
Keyur Shah, R.S. Besser

Session 132 - Combustion - Nox and Sox Control and Modeling

CoChair: Virendra K. Mathur

Chair: Paul Tripathi

137 **Long-Term Sulphation Behavior of Limestone Sorbents**
Jinsheng Wang, Edward J. Anthony

138 **Novel Sorbents for Mercury Capture from Fuel Gas**
Evan J. Granite, Christina Myers, Dennis Stanko, Henry W. Pennline

139 **The Pco Process for Photochemical Removal of Mercury from Flue Gas**
Evan J. Granite, Henry W. Pennline, Christopher R. McLarnon

140 **Performance Optimization and Emission Reduction of a Baled Biomass Combustor**
Michael E. Harasek, Martin Miltner, Angela Miltner, Christian Jordan, Anton Friedl

141 **Modeling and Simulation of Diesel Particulate Filters**
Herbert DaCosta

Session 146 - Water Reuse Technologies in Industrial Processes - Opportunities and Challenges Tutorial

Chair: C.B. Panchal

142 **Desalination and Water Reuse Technologies in the Petroleum Refining Industry**
Tom Pankratz

143 **Waste Heat and Thermal Desalination**
John Tonner

Kinetic analysis of Boudouard reaction and its dependence on porous structure

Shabi Ulzama, *Shabi.Ulzama@student.uni-magdeburg.de, Institute of Fluid Dynamics and Thermodynamics, Otto-von-Guericke University Magdeburg, Germany-39016,* ***E.Specht***, *eckehard.specht@vst.uni-magdeburg.de, Institute of Fluid Dynamics and Thermodynamics, Otto-von-Guericke University Magdeburg, Germany-39016*

Abstract

This work includes experimental targets for the re-estimation of kinetic rate coefficient of Boudouard reaction, which depends on the type of coke. A new set of activation energies and pre exponential factors presented here differs mostly from the values available in the literature. The work also includes experimental determination of sorption coefficients of CO_2 and CO. Theoretical analyses are reported to describe the importance of Boudouard reaction during those combustion processes where the value of excess air number is approximately 1. The results of simulations are compared with experimental data available in the literature. The simulation suggests that Boudouard reaction has a great influence on the process and experimental results reported in the present study show its dependence on the type of coke used in the process.

Introduction

The Boudouard reaction ($C+CO_2 \rightarrow 2CO$) has technical importance in many chemical engineering processes, being highly endothermic, and consuming carbon directly from the coke in many metallurgical processes. For example, the blast furnace process to reduce iron ore [1], the cupola furnace process for melting iron scrap, the shaft kiln process to produce lime and dolomite ($CaCO_3 \rightarrow CaO + CO_2$), the production process of manganese and chromium alloy [2] etc. Modeling of each of these processes can not be described mathematically without the knowledge of the kinetic parameters of Boudouard reaction and their dependence on different types of coke. A great number of modeling as well as experimental studies for better understanding of reaction mechanism and determination of reaction coefficient have been reported in the span of last few decades [3-7]. Consequently, it has been found that the values of kinetic parameters are dependent on the type of coke used for the process and influenced by the porous nature of the specimen. Generally, it is assumed that the Boudouard reaction can be described using the Langmuir-Hinshelwood formulation:

$$M_c = \frac{K_{Bou}.P_{CO_2}}{\left(1 + K_{CO_2}.P_{CO_2} + K_{CO}.P_{CO}\right).R.T}. \tag{1}$$

The existing experimental investigations [8-10] of Boudouard reaction show data discrepancy where the value of activation energy ranges within 100-300 kJ/mol and the values of reaction coefficient are 1 to 10 times higher than one another. In this work, the experimental study to explicate the data incongruity of the values of kinetic parameters and their dependence on coke types has been carried out in a tubular furnace. Moreover, this comprises also the mathematical modeling of special cases among the combustion processes such as the combustion chamber to control NO emissions; where O_2 concentration is not sufficient to complete the combustion (air excess number~1). During these processes, O_2 concentration keeps on decreasing because of combustion and it is counterbalanced by the production of CO_2. At higher conversion, the particle is exposed to the high concentration of CO_2 and it might be the only possible reason that there is always 100% conversion of particle even at non-existential concentration of O_2 at the end of the process. The kinetic analysis based on experiments provides the values of activation energy, the pre-exponential factor of the reaction and the sorption coefficients of CO_2 and CO. The new set of activation energies, pre exponential factor and particle conversion are compared against a variety of data available in the literature.

Experimental Descriptions

A re-estimation of kinetic rate constants of Boudouard reaction, which depends on the type of coke, has been carried out under a wide condition range (900-1200^{0}C). The details of the experimental setup are given in Fig. 1.

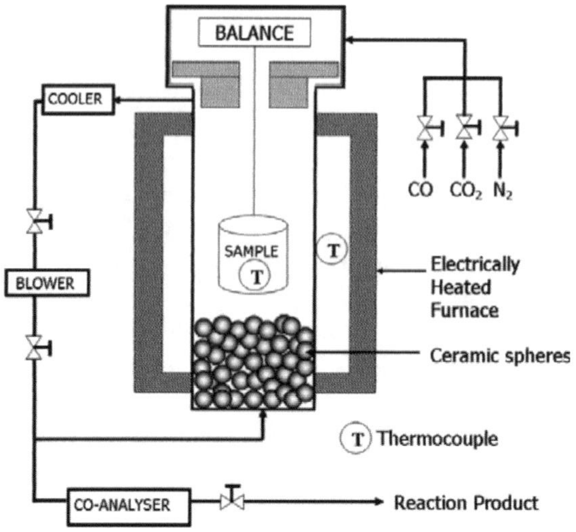

Figure 1: Schematic representation of the experimental setup.

Experiments were conducted inside a tube furnace having a diameter of 0.089 m and a height of 1.2 m, using an environment of CO_2-CO-N_2 gaseous mixture at 1 bar. A variety of cokes (Graphite, Poland coke, Czech coke) with different properties were taken for the experiments. The specimens used here have fixed geometry of a cylinder; with a diameter of 20 mm and a height of 50 mm, and of a plate; with a length of 90 mm, a width of 50mm and a thickness of 10 mm. The samples were placed on a balance which continuously measured the converted mass of carbon with time. The experiments covered the measurements of particle temperature, weight loss and mass transfer coefficient. Mass transfer calculations are needed to confirm that the process is mainly influenced by the chemical reaction only. Calculation of the values of mass transfer coefficient is based on the analogy between the heat and mass transfer. Aluminum plate with a length of 90 mm, a width of 50 mm and a thickness of 10mm was used to calculate the convective heat transfer rate from the hot plate surface to the CO_2 gas flowing at a lower temperature. Based on the assumption of the lumped parameter model i.e. the temperature of the whole body is same as the core temperature, the core temperature of the body was measured continually. The influence of mass transfer was reduced by adjusting the gas flow-rate and specimen dimensions. Consequently, it gives the true values of reaction coefficients. The higher gas flow rates require the gas to be circulated in a closed circuit. However, a part of the circulating gas is being taken off continuously to avoid high concentration of CO. In the lower section of the furnace, the gas flows through a fixed bed of inert ceramic spheres to attain a homogeneous flow over the cross section as well as temperature equal to that of the furnace. Moreover, an increase in gas velocity was limited to 3 m/s so that the mass loss from specimen surface, which is caused by removal of small particles due to high gas flow rate, could be prevented. With graphite slabs having a porosity of 1.2%, the diffusive transport of the reactant into the pores of a sample can be considered negligible. Thus, the outer surface of the graphite specimen can be assumed as the reaction surface and the value of reaction coefficient obtained can be taken as the true value. We have also performed a number of experiments using porous coke (Poland coke and Czech coke) to describe the influence of porosity on the reaction rate.

Model Description

The numerical modeling of spherically symmetric single particle behavior in an environment of CO_2-O_2 was performed by solving diffusive heat and mass transport equations coupled with chemical reaction in a shrinking coke. Effective values of process parameters i.e. diffusivity and thermal conductivity, are used to consider the effect of internal diffusion. The reaction chemistry of the combustion process can be described by the following heterogeneous reactions:

$$C + \frac{1}{2}O_2 \Rightarrow CO \tag{2}$$

$$C + CO_2 \Rightarrow 2CO \tag{3}$$

and one homogeneous reaction

$$CO + \frac{1}{2}O_2 \Rightarrow CO_2 . \tag{4}$$

However, the model does not take into account the homogeneous gas reaction inside the particle. It has been considered that the particle is preheated sufficiently so that the moisture content present inside the particle becomes negligible. Consequently, with a preheated particle it can be considered that CO does not get enough OH radicals to oxidize inside the porous body [11].

The model is based on following assumptions:

i. the coke particle is spherically symmetric,
ii. ash goes off from the particle surface during oxidation,
iii. pseudo steady state particle combustion.

The governing mass and energy conservation equations of the particle combustion model are:

$$\frac{1}{r^2}\frac{d}{dr}\left(D_{eff,i}.r^2\frac{dC_i}{dr}\right) - (k_i.O_m.\rho_c.C_i) = 0 \tag{5}$$

$$\frac{1}{r^2}\frac{d}{dr}\left(\lambda_{eff}.r^2\frac{dT}{dr}\right) - (k_i.O_m.\rho_c.C_i).\Delta H_{Ri} = 0 , \tag{6}$$

where C_i is the concentration (mol.m^{-3}) of species i (O_2, CO_2), $D_{eff,i}$ is the effective diffusion coefficient (m^2.s^{-1}), r is the radial position (m) from the particle center to the surface, O_m is the internal surface area (m^2.kg^{-1}), ρ_c is the density (kg.m^{-3}) of coke, k_i is the reaction coefficient for reaction between species i and carbon (m.s^{-1}), λ_{eff} is the effective thermal conductivity(J.m^{-1}.K^{-1}.s^{-1}), T is the particle temperature (K), and $\Delta H_{R,i}$ is the reaction enthalpy (kJ.mol^{-1}).

It has been found from experimental work done by many researchers that initially, the internal surface area increases with conversion, because of new pore openings and increase in pore diameter, to reach a maximum value and decreases henceforth. As combustion proceeds, the particle density decreases either due to a new pore opening or a change in pore diameter which leads to an increase in the gas content inside of particle despite of approximately no change in particle diameter. Furthermore, it has been reported [12] that the porosity of particle has an influence on effective diffusivity i.e. a more porous particle possesses a higher diffusivity. The overall diffusivity follows a decrease with conversion. Therefore, the initial values of total surface area and the diffusivity at room temperature can no longer be used for the complete simulation. As a first approximation to solve these coupled differential equations (Eqs. 5&6), the value of the ratio of product of internal surface and density to effective diffusivity is considered to be a constant. On the basis of experimental data, the appropriate values of these parameters were chosen to perform simulation.

Table 1: Equations for physical properties and process parameters

Parameter	Equation	Unit
Nusselt number	$Nu = 2 + 1.1 \times \mathrm{Re}^{0.50} \times \mathrm{Pr}^{0.33}$	[-]
Sherwood number	$Sh = 2 + 1.1 \times \mathrm{Re}^{0.50} \times Sc^{0.33}$	[-]
Effective thermal conductivity	$\lambda_{eff} = 0.23 + 2.44.10^{6}.(T - 273)^{1.7}$	$J.m^{-1}.K^{-1}.s^{-1}$
Mass transfer coefficient	$\frac{Sh.D_g}{2.R}$	$m.s^{-1}$
Heat transfer coefficient	$\frac{Nu.\lambda_g}{2.R}$	$J.m^{-2}.K^{-1}.s^{-1}$
Particle diffusivity	$D_p = 20\exp\left(-\frac{140kJ/mole}{R.T}\right)$	$m^2.s^{-1}$
Effective diffusivity	$\frac{1}{D_{eff}} = \frac{1}{D_p} + \frac{1}{\left(\frac{D_g \varepsilon}{\tau}\right)}$	$m^2.s^{-1}$
Reaction coefficient (Boudouard reaction)	$K = 1.428 \times 10^6 \exp\left(-\frac{226kJ/mole}{R.T}\right)$	$m.s^{-1}$
Reaction coefficient (carbon oxidation reaction)	$K = 1.1 \times 10^4 \exp\left(-\frac{180kJ/mole}{R.T}\right)$	$m.s^{-1}$

The boundary conditions included application of zero gradients at the center of the particle,

$$\frac{dC_i}{dr} = 0, \quad \frac{dT}{dr} = 0, \tag{7}$$

and conditions at the particle surface,

$$\frac{dC_i}{dr} = \frac{\beta_i}{D_{eff,i}}\left(C_{\infty,i} - C_{s,i}\right), \tag{8}$$

$$\frac{dT}{dr} = \frac{h}{\lambda_{eff}}(T_\infty - T_s) + \frac{e \times \sigma \times \left(T_\infty^4 - T_s^4\right)}{\lambda_{eff}}. \tag{9}$$

where β_i is the mass transfer coefficient ($m.sec^{-1}$), $C_{\infty,i}$ and $C_{s,i}$ are the concentrations ($mol.m^{-3}$) of species *i* in the ambience and at the particle surface respectively, *h* is the heat transfer coefficient ($J.m^{-2}.K^{-1}.s^{-1}$), T_∞ and T_s are the temperatures (K) in the ambience and at the particle surface respectively.

Overall reaction coefficient can be calculated by the following equation, which includes both the mass transfer effect and the chemical reaction,

$$\frac{1}{k_{overall,i}} = \frac{1}{\beta_i} + \frac{1}{\sqrt{k_i.O_m.\rho_c.D_p}} \tag{10}$$

Particle conversion can be calculated by the following equation,

$$x = \left\{1 - \left(\frac{r_t}{r_0}\right)^3\right\}, \tag{11}$$

where the value of r_t can be calculated using the following equation,

$$r_t = \bar{r} - \frac{0.12.K.C_{overall}.t}{\rho}, \tag{12}$$

here $\bar{r}$ denotes the radius of the particle after each simulation run. The values of parameters used in simulation are given in Table 2.

Table 2: Parameters used in simulation

Parameter	Value	Unit
T_∞	1123	K
Tortuosity (τ)	5	[-]
O_m	500	$m^2.kg^{-1}$
ΔH_{O_2}	-396	$kJ.mol^{-1}$
ΔH_{CO_2}	172.5	$kJ.mol^{-1}$
Emissivity (e)	0.85	[-]

Results and Discussion

Calculation of mass transfer coefficient

Analogy between the heat and mass transfer based on experimental findings using aluminum specimen is used to calculate the values of mass transfer coefficient. Figure 2(a) shows the measured temperature-time plots of the core of aluminum plate with different volumetric flow-rate of CO_2. Specimens were kept at various initial temperatures. Experiments with aluminum plate were needed to confirm that there is no influence of reaction as it is in the case of graphite slab.

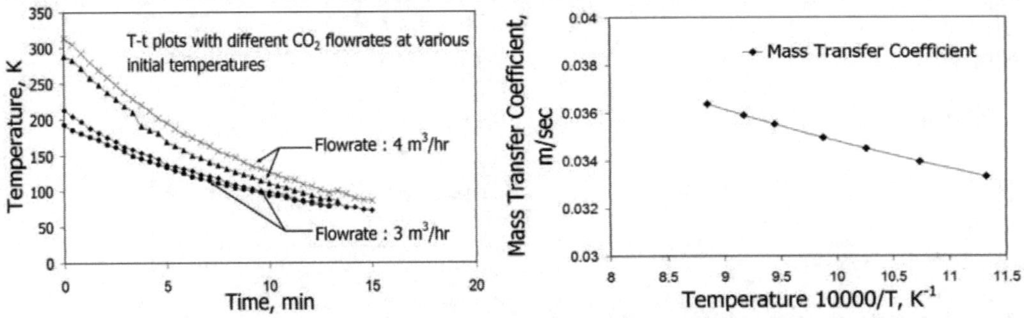

Figure 2: (a) Temperature-time plot of aluminum plate with different volumetric flow rates of CO_2. **(b)** Variation of mass transfer coefficient with temperature.

These experimental findings of temperature variations were used to calculate the values of heat transfer coefficient. Afterward, the values of mass transfer coefficients were calculated based on the measured values of heat transfer coefficients. Figure 2(b) describes the variation of mass transfer coefficient with temperature. Experimentally found values of mass transfer coefficient confirm that there is no significant influence of mass transfer on the process and the process is mainly influenced by reaction only. Moreover, it is further investigated by changing the value of flow rate at a particular temperature and it has been found that the influence of mass transfer on the process is not significant.

Calculation of true reaction coefficient of Boudouard reaction and its dependence on porosity

A graphite slab with dimensions described previously in experimental description has been used to calculate the true value of reaction coefficient. Since graphite has porosity of 1.2 %, it is assumed that there is no significant contribution of pore diffusion over the process. On this basis, we can assume that the loss in specimen weight is only because of reaction taking place at the outer surface. The pure CO_2 was used for the experimental runs at different temperatures in the range of 900 to 1200 ^{0}C. It has been found that a linear variation of conversion with time (weight loss relative to the time and the outer geometrical surface) exists. Figure 3 illustrates a summary of the experiments carried out at different temperatures. The values of reaction coefficients at different temperatures can be easily calculated from the slope of linear conversion vs. time plot.

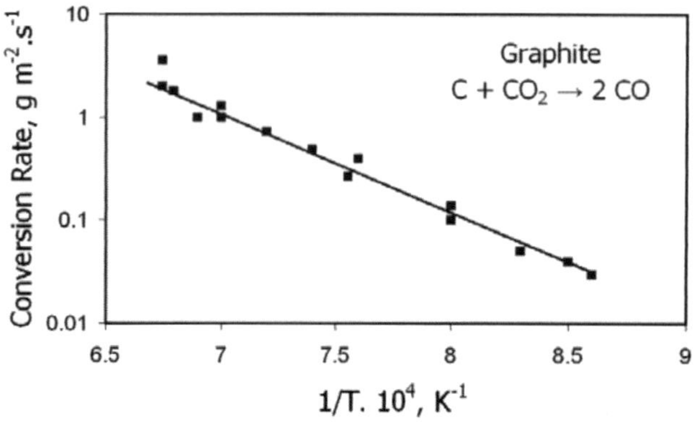

Figure 3: Conversion rate of graphite as a function of the temperature in an environment of CO_2

These measured values of reaction coefficients are plotted against the temperatures to calculate the unknown values of coefficients of Arrhenius equation. The value of activation energy is compared against the data available in the literature (Table 3).

Table 3: Comparison of experimental values with values found in the literature

Reference	Activation Energy (kJ.mol^{-1})
Walker, Foresti and Wright	201
Specht and Jeschar	222
Beyer, Pückoff and Ulrich	227
Agrawal and Sears	223
Seidel	305
Present experimental value	226

It can be seen from Table 3 that the experimental value of activation energy is in agreement with other experimental values calculated by different researchers.

Figure 4 shows the comparison of experimental values of reaction coefficient with other researchers (semi log plot). As shown in Fig. 4, one of the possible reasons of the experimental values differing from the values obtained by other studies is the way how the rate coefficients are defined. However, many of the studies define the rate coefficient considering the sorption coefficients [9]. We have defined our reaction coefficient while considering the values of sorption coefficients equal to zero. Nevertheless, it can be found that an assumption of the sorption coefficients to be equal to zero would cause the true value of reaction coefficient (from the literature) to decrease by a factor of ~5 i.e. in the range of present experimental work. On the basis of experimental investigation, Arrhenius equation can be described as follows;

$$K_{Bou} = 1.428 \times 10^{6}.\exp\left(-\frac{226kJ/mole}{R.T}\right) \quad (13)$$

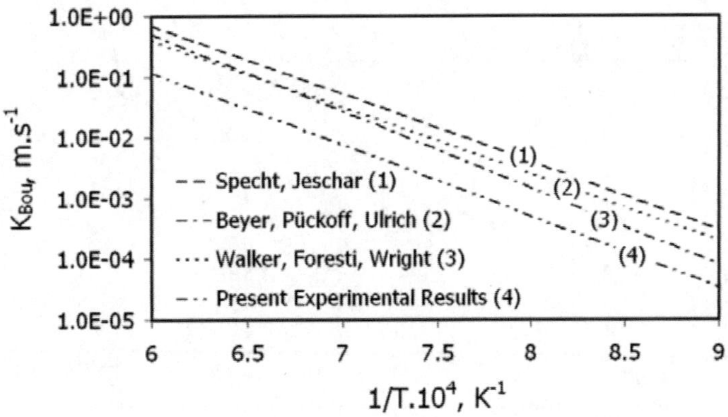

Figure 4: Comparison of experimental reaction coefficient with the values found in the literature.

The apparent values of activation energy obtained while considering the porous nature of the coke are illustrated in Table 4. It describes the dependence of Boudouard reaction on porosity of different types of coke. During the process, the internal surface area changes and it is difficult to say what the value it is at a particular time. However, in one of the experimental investigations it has been found that the apparent values are much lower that the true value of activation energy and exhibit a lower temperature dependency.

Table 4: Apparent and true activation energy of Boudouard reaction

$C+CO_2 \rightarrow 2CO$	
Coal Type	**Activation Energy (kJ mol^{-1})**
Graphite	226
Poland coke	166
Czech coke	141

Experimental measurement of sorption coefficients of CO_2 and CO

To investigate individually the influence of the CO_2 and CO over the rate of Boudouard reaction, the experiments were carried out in mixture of: CO_2-N_2 or CO_2-CO-N_2; while maintaining the temperature constant. Equation 1 can be rewritten as follows;

$$\frac{M_c}{R.T}.\frac{P_{CO_2}}{\dot{m}_c} = \frac{1}{K_{Bou}} + \frac{K_{CO_2}}{K_{Bou}}.P_{CO_2} \quad (14)$$

$$\frac{M_c}{R.T}.\frac{K_{Bou}.P_{CO_2}}{\dot{m}_c} - 1 - K_{CO_2}.P_{CO_2} = K_{CO}.P_{CO} \quad (15)$$

Experimentally found values have been plotted in accordance to the Equations (14) & (15) in succession to calculate the values of sorption coefficients of CO_2 and CO. Figure 5 describes the behaviors of Equations (14) & (15) for a temperature value of 1000 ^{0}C. As one of the experimental outcomes, it has been found that the values of sorption coefficients K_{CO_2} ~ 2.4 bar^{-1} and K_{CO} ~ 7.0 bar^{-1} for CO_2 and CO respectively. From Fig. 5, we can see, though the CO profile doesn't follow a linear dependency with partial pressure of CO, an approximation of CO sorption coefficient while considering its linear profile gives a reasonable value. Dependence of both CO_2 and CO on temperature will be published in further studies.

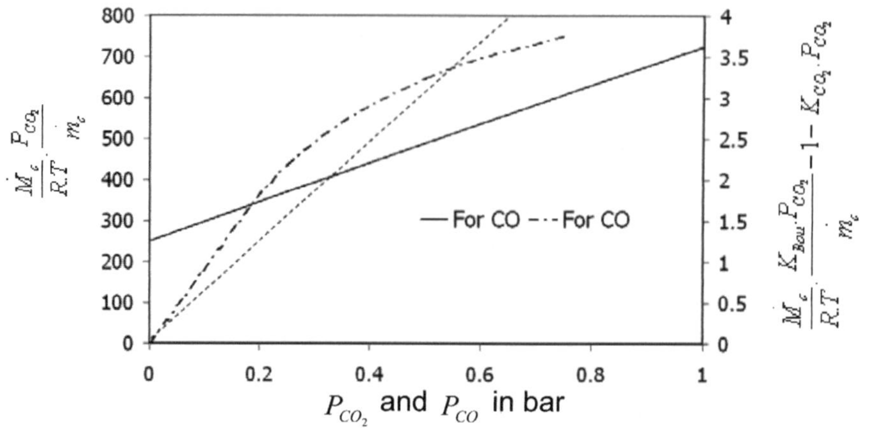

Figure 5: Measured experimental values for calculation of the sorption coefficients of CO_2 and CO (--- denotes the approximation).

For the lower values of CO_2 partial pressure, the product of $K_{CO_2}.P_{CO_2}$ has a lower value compared to 1. This might be one of the reasons why the previous studies carried out in an environment of lower values of CO_2 partial pressure were not able to confirm Langmuir-Hinshelwood formulation for the Boudouard reaction.

Fractional conversion and particle shrinkage

Experimentally found value of reaction coefficient is used in steady state solution of the model. The Equations (5) & (6) are complex coupled differential equations, where the process parameters i.e. diffusivity, reaction coefficients and effective thermal conductivity; are strongly influenced by the temperature. Figure 6 shows comparison of model predictions to the data available in the literature for fractional conversion of a single particle in infinite volume of O_2 and CO_2.

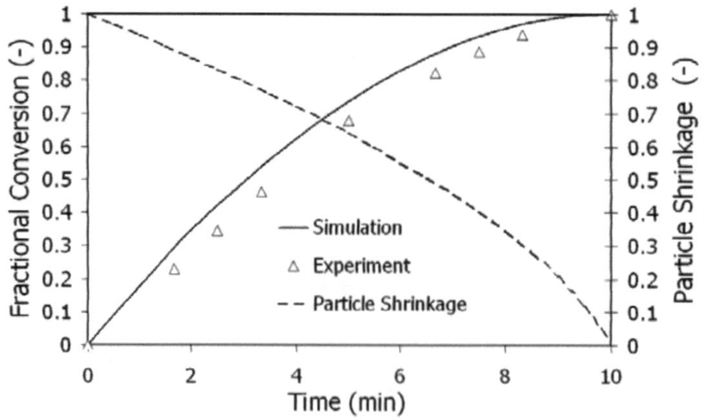

Figure 6: History of fractional conversion and particle shrinkage with time at ambient temperature 1123 K with 12 vol% O_2 and 10 vol% CO_2.

Specimen used for experiment was from Poland coke with process conditions of 12 vol% O_2 and 10 vol% CO_2 at temperature 1123 K. One of the major problems associated with single particle combustion is the experimental measurement of particle shrinkage (r_t/r_0) with time. However, the simulation is able to describe particle shrinkage with time (Fig. 6). The model simulation results show a good agreement with experiment.

Conclusion

This study provided the experimental measurement of reaction coefficient of Boudouard reaction. Experimentally found rate expression corresponds to a group of values proposed by different studies. Experimental results obtained have shown a strong dependence of reaction coefficient on the types of the coke used. However, the apparent values are much lower than the true values and exhibit a lower temperature dependency. Moreover, the values of the sorption coefficients are successfully determined. However, the dependencies of these coefficients on temperature will be considered in further research. Finally, a goal of this work was to demonstrate a steady state model that accurately describes single particle behavior in an environment of CO_2-O_2. The comprehensive kinetic relation of Boudouard reaction obtained experimentally provides a satisfactory description of the process over the entire range of conversion. Nevertheless, the present model is sufficient enough to describe fundamental numerical computations.

Acknowledgement- The authors gratefully acknowledge the support from the Deutsche Forschungsgemeinschaft (DFG).

References

1. Grabke H.J., Krajak R. and Nava Paz J.C. (1993), "On the mechanism of catastrophic carburization: 'metal dusting' ", Corrosion Sci. 35, pp.1141-1145.
2. Kaczorowski J. (2003), "The Boudouard Reaction in Mn-alloy Production Processes", International seminar in materials processing, November.
3. Turkdogan E.T. and Vinters J.V. (1969), "Kinetics of Oxidation of Graphite and Charcoal in Carbon dioxide", Carbon, vol.7, pp.101-117.
4. Alam M. and Debroy T. (1987), "A Comparative Study of the Roles of KCN and NaCN as Catalytic Precursors in the Boudouard Reaction", Fuel, vol.66, pp.103-112.
5. Cheng H., Reiser B.D. and Dean S. (1999), "On the Mechanism and Energies of Boudouard Reaction at FeO (1 0 0) Surface: $2CO \rightarrow C + CO_2$", Catalysis Today, vol.50, pp.579-588.
6. Grabke H.J. (1980), "Oxygen Transfer and Carbon Gasification in the Reaction of Different Carbons with CO_2", Carbon, vol.10, pp.587-599.
7. Laurendeau N.M. (1978),"Heterogeneous Kinetics of Coal Char Gasification and Combustion", Progress in Energy and Combustion Science, vol.4, pp.221-270.
8. Agrawal A.K. and Sears J.T. (1980), "The Coal Char Reactions with CO_2-CO Gas Mixture", Ind. Eng. Chem. Process. Des. Dev. 19, pp.364-371.
9. Specht E. and Jeschar R. (1986),"Reaktionsmechanismus und Abrandgeschwindigkeit mit Luft angeströmter Kohlenstoffteilchen", Steel Research 57, pp.148-159.
10. Walker P.L., Foresti R.J. and Wright C.C. (1953), "Surface Area Studies of Carbon-Carbon Dioxide Reactions", Ind. and Eng. Chem. 45, pp.1703-1710.

11. Howard, J.B.; Williams, G.C.; Fine, D.H., 1973, Kinetics of carbon monoxide oxidation in postflame gases, *14th Symposium (International) on Combustion*, 975-986.
12. Wang, H.; Dlugogorski, B.Z.; Kennedy, E.M., 1999, Theoretical analysis of reaction regimes in low-temperature oxidation of coal, *Fuel*, 78, 1073-1081.

Paper 27b

Two-Phase Pipe Side Process Model for Fired Heaters

C.B. Panchal **, Steve Lottes, and Michael Petrick
Argonne National Laboratory
9700 S Cass Avenue
Argonne, IL 60439
CPanchal@anl.gov

** To whom correspondence should be sent

Key Words: fired heater, crude heater, two-phase pipe flow, petroleum fouling, API530

Prepared for Presentation at the 2006 National Meeting, Orlando, FL, April 23-27, 2006

Two-Phase Pipe Side Process Model for Fired Heaters

C.B. Panchal, S,A. Lottes, and M. Petrick

Argonne National Laboratory

Argonne, IL 60439

ABSTRACT

The objective of this project was to develop multiphase heat transfer and pipe flow analytical methodologies that are applicable to large diameter pipes and U-bends found in refinery heaters. The technical approach was implemented in three phases. Phase I focused on identifying and evaluating multiphase correlations to predict process side flow regimes, pressure drop, and the heat transfer coefficient applicable to fired heaters. This involved an extensive literature survey, systematic evaluation of literature correlations for applicability to process heaters, and selection of the most promising correlations for developing and comparative validation of the process model in Phase II. Phase III is being pursued to incorporate the Argonne's threshold fouling model into the process model.

INTRODUCTION

Process heater is a work horse of refining and petrochemical industries. A typical refinery consumes about 445,000 Btu/barrel of crude processed, most of which is provided by process heaters. The recent trend of high utilization in the refining industry enforces operating of process heaters close to, or beyond, design capacities. This requires better understanding of the process side two-phase flows and predictive methods. Argonne is pursuing a long-term project for developing a pipe-flow process model, which includes multiphase flow regime predictions, heat transfer, pressure drop in straight sections of the heater pipes and U-Bends. The process model will then be expanded to include a validated coking model to determine threshold coking conditions. Argonne joined an on-going DOE project whose objective was to develop a new generation of refinery heaters that are highly energy efficient and have low NOx emissions. The technical approach was implemented in two phases. The objective of Phase I effort was to identify and evaluate multiphase literature correlations to predict process side flow regimes, pressure drops, and heat transfer coefficients. This involved an extensive literature survey, systematic evaluation of literature correlations for applicability to refinery heaters, and selection of the most promising correlations. The Phase II tasks included 1) the development of a process model; 2) comparative analysis of the literature correlations to select those applicable to fired heaters; and 3) comparative analysis of the process model with the API-530 correlation. In the next phase a fouling/coking model is being incorporated into the process model to determine the localized coking rate and threshold coking conditions. In the present analysis this process model is applied to a generic crude heater.

EVALUATION OF LITERATURE CORRELATIONS

A broad spectrum of multiphase flows are encountered within the process side tubes in process heaters as the fluid is vaporized. The multiphase flows that occur are the result of the heater design and operating parameters. The flow regime (characterized by the distribution of vapor and liquid) changes dramatically along the pipe length and thus directly impacts the heat transfer and pressure drop. In order to design and safely operate the fired heaters appropriate heat transfer and pressure drop models/correlations must be utilized.

A critical review of the literature data and prediction correlations was performed to identify literature correlations that could be applicable for fired heaters. The bases for selecting correlations from this initial screening analysis were as follows:

- Most frequently cited in the literature;
- Recent literature reviews and comparison with the literature data; and
- Based on phenomenological models which incorporate a) physical properties, b) pipe size, c) orientation (horizontal vs. vertical up and down flows; d) individual flow regimes.

A comparative analysis was performed to determine applicability of these correlations to process conditions typical of refinery fired heaters. In general, the literature correlations have been developed with experimental data obtained under test conditions that were outside (sometimes significantly) the parameter space that is typical for refinery heater conditions. A key objective of the analysis was to identify correlations that are based on mechanistic models and/or wide range of experimental databases to minimize the prediction uncertainty in applying to refinery heaters.

The outcome of the comparative analysis was a list of recommended correlations that can be used to analyze fired heaters; they are presented in Table 1 for straight sections and in Table 2 for U-Bends. These correlations were subsequently incorporated into the performance prediction model for analyzing process heaters.

Table 1 Recommended Correlations for Straight Sections

	Horizontal Flows	Vertical Flows	
		UP Flows	Down Flows
Flow regimes	Taitel-Dukler (1976)	Speddings et al. (1998) Hewitt-Robert (1969)	Speddings et al. (1980)
Pressure drop	Chisholm (1973)	Slug/churn Flows – Friedel (1976) Annular - Speddings et al. (1998)	Slug/churn Flows – Friedel (1976) Annular - Speddings et al (1998)
Heat transfer coefficient	Slug flow – Shah (1976) Annular – Chen (1966)	Annular – Steiner&Taborek (1992) Slug/Churn – Wadekar (1990)	Annular – Numerich-Claus (1991) Slug/Annular - Wadekar (1990)
Void fraction		Spedding-Hand (1995)	Spedding-Hand (1995)

Table 2 Recommended Correlations for U-Bends

	Horizontal Bends	Vertical Bends
Flow regimes		Chen et al. (1983)
Pressure drop	Chisholm (1980)	Chisholm (1980)
Heat transfer coefficient	Farukhi (1973) Moshfeghian (1978)	Annular – Farukhi (1973); Moshfeghian (1978) Stratified – Katan (1998)

PROCESS MODEL

The technical approach taken in developing a fired heater process model was to incorporate the selected literature correlations and calculation methods into a Fortran-based model for predicting local process parameters as well as the overall performance of the fired heater. The process model predicts the performance of both the convective section, including the steam coil, and the radiative section. The performance analysis of the convective section is analyzed first, taking process conditions at the cross-over point as input parameters. Point-wise integration is followed for each row to predict process conditions at coil inlet and the heat rate of the superheated steam section. The process model predicts the stack-gas temperature and can be compared with the measured value. Due to cross-flow arrangement of process and steam pipes with respect to flue-gas flow, temperature of the stack is expected to be non-uniform. Depending on location of temperature sensor, the measured stack gas temperature may or may not be representative of a mixed-average value. The process model also predicts stack-gas pressure drop across the convective section. The bridgewall temperature can be derived from the combustion model. Alternatively, if accurate measurement of stack-gas temperature is available, it can be used as an input parameter. An iteration procedure is then used to calculate the bridgewall temperature.

Process Model for Radiative Section

The radiative section process model predicts flow-regime based heat transfer coefficients and pressure drops. This approach is based on an assumption of uniform flow distribution in all passes. However, if maldistribution of flow is suspected, the process model can be applied to individual passes separately. The model was unidirectionally coupled with the specified heat-flux distribution to calculate process conditions along each pipe of a given pass. The computed parameters include flow rates of vapor and liquid phases, axial pressure profile, and pipe wall temperatures. The process model computes local conditions at each step and selects correlations based on flow regimes. The forward-march computation starts at the crossover point for one of the passes and ends at coil-out point. The Fortran program has been written with separate subroutines for calculating flow regimes, pressure drops, and heat transfer coefficients based on local process conditions. The computer model is structured so that, if desired, it can be readily incorporated into a comprehensive fully coupled Fired Heater model. The results from the Fortran program are transferred into an Excel™ program for post processing the results.

Performance Model for Convective Section

In order to simulate the overall performance of the fired heater, it was necessary to develop a performance model for the convective section. In addition to process side calculations, the performance model predicts the flue-gas side heat transfer coefficient and pressure drop. Analysis of the cross-flow between flue gas and process fluid requires a 2-dimensional calculation method. A forward-march computation method is employed starting from the bridgewall point where the flue gas enters the convective section. The computation of local heat flux and pressure drop is carried out for each row along with the flue-gas temperature and pressure drop. At the end of this computation, the coil-in process temperature is compared with the measured value and iteration is performed to match these values. If the stack-gas temperature is used as the input parameter, the iteration is on the bridgewall temperature.

Because no assumptions are made for calculating the local heat flux, this first principle performance prediction method provides an independent check on the heat duty in the convective section and hence for the whole heater. Fluid process conditions, such as temperature, vapor quality, and pressure, at crossover are generally not always measured and thus must be computed; these values are needed as input parameters for the radiative section model.

Coking/Fouling Model

Panchal et al (1997) developed a threshold fouling model applicable to preheat train. The model was developed for single-phase flows; however, it is applicable to two phase flows provided local fluid dynamics can be calculated (Kuru and Panchal, 1997). This fouling model was incorporated into this process heater model. The rate of fouling is expressed in terms of generation of precursor in the boundary layer and its removal before firm adhesion of deposits on the wall occurs.

$$dR_f/dt = \alpha \, Re^{\beta} \, Pr^{-0.33} \exp^{(-E/RT_f)} - \lambda \, \tau_s$$

Where Reynolds number, Re and Prandtl number, Pr are calculated for properties at the bulk and film temperature T_f (average of bulk and metal wall temperatures), respectively. The Reynolds number in this equation was based on the two-phase flow viscosity. The wall shear stress, τ_s is calculated for the liquid flow at the wall surface. Constants α, β, λ and the activation energy E are determined from laboratory and/or field data. When the equation is equated to zero, it provides threshold fouling/coking conditions between the in-side tube metal temperature and the wall shear stress.

CRUDE HEATER

The two-phase process model was applied to a crude heater for which overall operating conditions have been reported by Lawler (1979). In a previous study, the effects of preheat train fouling on energy consumption were analyzed for this refinery (Panchal and HuangFu, 2000). Operating conditions used in this model are summarized in Table 3. Heater dimensions were not given by Lawler (1979). A heater with four passes, 8-inch pipe diameter (7.625 inside diameter), and 30 vertical pipes per pass was modeled in this analysis. With this heater configuration the mass velocity was 278 lb/s ft^2. Physical and thermodynamic properties of typical medium crudes with API gravity in the range of 30 and 35 were used.

Table 3 Process parameters of the crude heater for the process model

Process Parameters	Input	Calculated
Process Fluid	Crude	
Crude Flow Rate, lb/hr	1,271,325	
Mass Flow Rate, lb/s ft^2 (for assumed pipe size)	278	
Vapor Quality at Coil Inlet	0.05	
at Crossover		0.15
at Coil Outlet		0.47
Coil Inlet Pressure, psia		406
Pressure at Crossover, psia		178
Coil Outlet Pressure, psia (assumed)	30	
Temperatures, °F		
Coil In (for clean preheat train conditions)	560	
Crossver		597
Coil Outlet		709
Bridgewall (assumed)	1,600	
Exit Flue Gas		569
Heat Rate Process, Million Btu/hr		
Convective Section		55
Radiative Section		146
Total		201
Average Radiation Heat Flux, Btu/hr ft^2		14,074

Flow Regimes

The process model predicts flow regimes at each point along the pipe length and selects appropriate correlations. The flow regime in the convective section is generally slug flow due to low vapor qualities. In the radiative section, the flow regime in up flow pipes starts with churn flow at low vapor qualities and then switches to annular flow as vapor qualities increases. However, the annular flow may not be fully developed and liquid slugs at some frequency would be expected. The flow regime in down flow pipe for most of the heater remains annular/slug, which is combination of annular flow with systematic slug flow at a regular frequency. The flow regime in U-bends of the radiative section at low vapor qualities is either separated or semi-annular, which could produce dry out on inner surface of U-bends and downstream regions. Later, the flow regime becomes annular, which would reduce potential dry out of the heated surface. An important fact to remember is that the transition of flow regimes could occur with change in process conditions, especially mass velocity and local vapor qualities. Therefore, evaluating the impact of possible change in flow regimes is always recommended.

Heat Transfer Coefficient and TMTs

The local heat transfer coefficients predicted by the process model and API-530 correlation differ significantly. Figure 1 shows that the heat transfer coefficient predicted by the process model increases as the vapor quality increases from crossover to coil outlet; however, no significant variation in the coefficient is predicted by API-530. Also, the heat transfer coefficients for up flow and down flow pipes are different, especially at lower vapor qualities. Note that API-530 correlation is independent of flow direction, up flow or down flow.

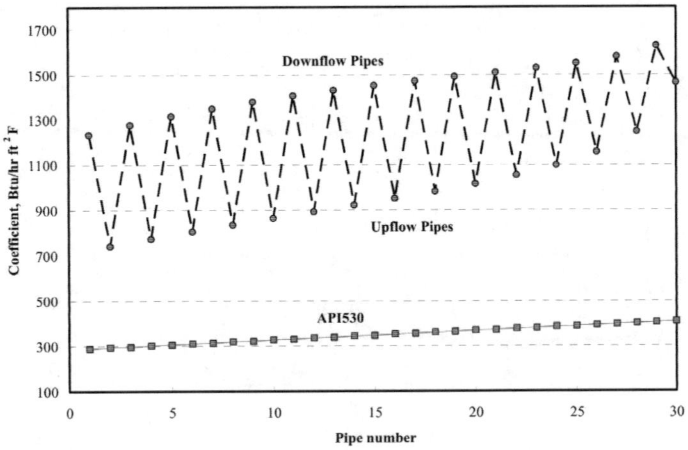

Figure 1. Heat transfer coefficient predicted by the process model and API-530.

Figure 2 shows TMTs based on circumferentially averaged heat fluxes for hot-end (Pipe#30 up flow) pipe with a higher vapor quality of 45% and cold-end pipe (Pipe#1 down flow) with a lower vapor quality of 15%. In general API-530 predicts higher TMTs at all heat fluxes, and the over prediction increases in the region of high heat fluxes. The maximum TMT predicted by API-530 correlation for Pipe#30 is 822°F as compared to 777°F by the process model. The difference between TMT predicted by the two methods at the point of TMT measurement would be even greater when circumferential variations of heat flux are taken into account. Note that the API-530 does not differentiate TMTs in up flow or down flow, while the process model shows a difference in expected TMTs for up flow and down flow pipes. Depending on flow regimes, and transition that may occur during change in process conditions, the maximum TMT may or may not occur at the measurement location or even on the pipe equipped with the TMT sensor.

Coking Rate

Constants used for analyzing fouling of the preheat train (Panchal and HuangFu, 2000) were used in this analysis. Figure 3 shows the coking rate ((hr ft^2 F/Btu)/day) for cold-end, intermediate, and hot-end pipes of the heater. The coking rate for the cold-end pipe is nearly ¼ the fouling rate calculated for the hot-end of the preheat train, which reflects the effects of high two-phase flow wall shear stress in the process heater as compared to single-phase flow shear stress in the preheat train exchanger on the local fouling rate. The coking rate increases with increase in crude and in-side tube metal temperatures; however, the increasing wall shear stress as the vapor quality increases has opposite effects. The coking rate accelerates towards the hot-end pipes of the heater. It should be noted that the coking rate is calculated based on circumferentially averaged heat flux and in-side tube metal temperatures. The coking would be much higher locally for circumferential variation of heat flux combined with two-phase maldistribution of liquid and vapor phases. Locations of individual pipes with respect to the burner configurations would further enhance this effect of localized coking. This process model is a first step towards predicting localized coking and taking corrective steps before significant coking occurs.

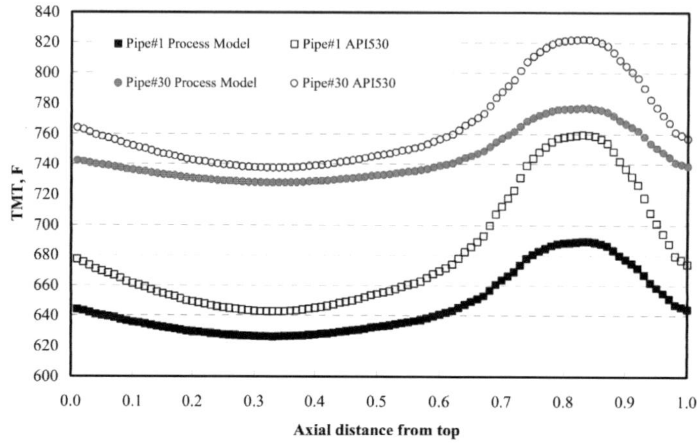

Figure 2. TMTs predicted by the process model and API-530.

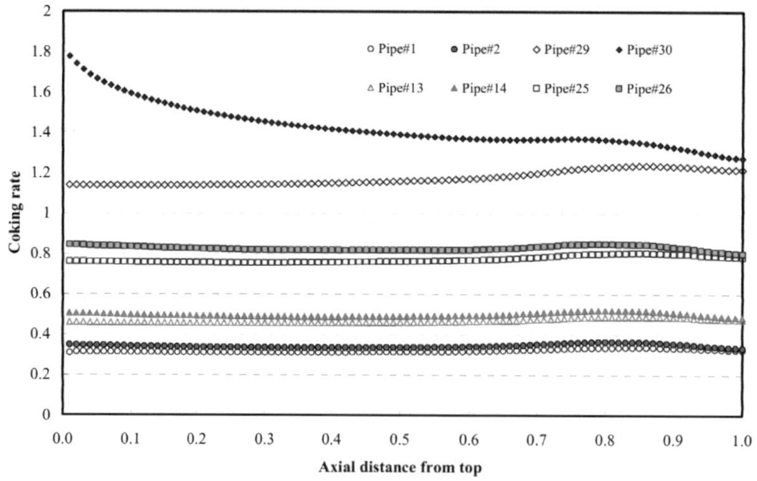

Figure 3. Coking rate (hr ft^2 F/Btu (10^{-4}) per day) distribution.

ANALYSIS OF U-BENDS

A considerable body of evidence exists in the literature that indicates that U-bends (that exist in process heaters) can have a major impact on the downstream fluid flow, heat transfer, and coking characteristics for distances up to 10-20 L/D or greater. This in turn could impact the location and magnitude of the TMT and thus the heater's safe operating range. An initial CFD study was therefore carried out to assess the major inertial effects on a two phase annular flow entering vertical U-bends for both up and down flow cases. While tube walls may be wet around the entire circumference in an annular flow entering a U-bend, the larger inertia of the liquid portion of the flow tends to force the liquid to the outer wall of the bend as the flow proceeds through the bend. A primary concern with this type of flow redistribution in fired heaters is that a region of wall dry out can be created on the inner wall of the exit portion of the bend. Dry wall regions (steady or oscillatory) are places where significant coke buildup generally occurs.

Test cases for the entire range of inlet flow rates showed regions of likely dry out on the inside wall at the exit of the bend in both up flow and down flow cases. The bulk of the mass of the liquid accumulates on the outside wall of the turn through the bend. Figure 4A shows the liquid-vapor distribution at a cross section of pipe two diameters up from the exit of a down flow U-bend. The mass flux in this case is 340 lb/s ft^2. The velocity vectors in the figure show the secondary flow pattern. The liquid occupies the dark region on the outside wall at the U-bend exit. Figure 4B shows the liquid distribution in the U-bend at a cut through its mid-plane. The liquid film on the wall at the entry is a little over one tenth of an inch thick and carries over 85 percent of the mass flow. A large accumulation of liquid in the bottom portion of the bend can be seen as well as an absence of liquid on the inner wall at the U-bend exit.

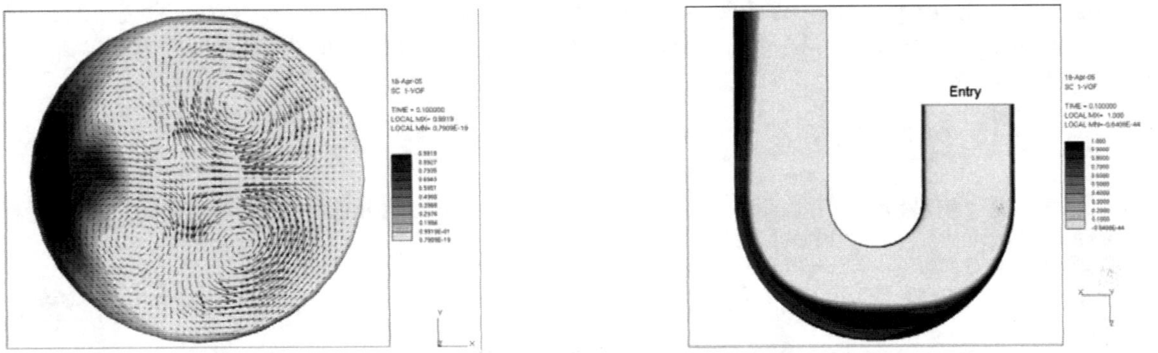

A B

Figure 4. A. Cross section at up flow exit of a down flow U-bend showing secondary flow patterns and accumulation of liquid (dark region) on the outer wall of the bend. B. Liquid distribution at mid-plane of a down flow U-bend.

CONCLUSIONS AND RECOMMENDATIONS

The major conclusions and recommendations from this analysis are as follows:

Conclusions:

- The literature survey showed there are new correlations and calculation methods for pipe sizes typical of refinery heaters.
- Prediction correlations, especially for heat transfer coefficient, are flow-regime dependent. A single correlation cannot be applied to all flow regimes in horizontal and vertical (up flow and down flow) pipes.
- Analysis of the crude heater shows that API-530 under predicts the process side heat transfer coefficient and hence over predicts TMTs.

Recommendations:

- This process model should be validated with plant data from refinery heaters.
- Considerable heat flux variations exist in the high heat-flux zone; this signifies the need for a combined model to determine the interactive effects of localized coking and TMTs.
- A considerable body of evidence exists in the literature and operating experiences that indicates U-Bend can have a major impact on the downstream fluid flow and heat transfer characteristics. Predictive methods are needed for circumferential variations of heat transfer coefficient.
- This process model focused on low exit vapor qualities typical of resid and crude heaters. It should be extended to high vapor qualities encountered in hydrotreating and reforming units, where regions of transition from annular to dispersed flow could have high propensity of localized coking.

ACKNOWLEDGEMENT

This project was funded by the Combustion Program of the DOE Office Energy Efficiency and Renewable Energy, the Office of Industrial Technologies Program. The authors acknowledge the support of Mr. R. Gemmer and Mr. J. Quinn of the Office of Industrial Technologies Program.

REFERENCES

Chen JC. 1966. A correlation for boiling heat transfer to saturated fluids in convective flow. I & EC Process Design and Development. 5, 322-329.

Chen TK et al., 1983. Flow pattern transition for gas-liquid two-phase flow in the bend of vertical hair-pin tubes. Proceedings of the third multi-phase flow and heat transfer symposium workshop. Miami, FL. Paper A3.6.

Chisholm D. 1973. Pressure gradients due to friction during the flow of evaporating two-phase mixtures in smooth tubes and channels. *Int J Heat Mass Transfer*. 16, 347.

Chisholm D. 1980. Two-phase flow in bends. *Int J Multiphase Flow*. 6, 361-367.

Farukhi MN. 1973. Two-phase flows in U-bends. *PhD Thesis*. Oklahoma State University, Stillwater, OK.

Friedel. 1979. New friction pressure drop correlations for upward, horizontal and downward two-phase pipe flow. *Presented at the HTFS Symposium*, Oxford, Hoechst AG Reference 372217/24 698.

Hewitt GF and Roberts DN. 1969. Studies of two-phase flow pattern by simultaneous X-ray and flash photography. *AERE-M 2159*.

Katan et al. 1998. Flow boiling in horizontal tubes: Part 1 – Development of diabatic two-phase flow pattern map; Part 2 – New heat transfer data for five refrigerants. Trans of the ASME. 120, 140-155.

Lawler D. 1979. Fouling of crude oil in refinery heat exchangers. *Proc. Conf on Fouling Science or Art*, The Institute of Chemical Engineers, UK. 155-165.

Moshfeghian M. 1978. Fluid flow and heat transfer in U-Bends. *MS Dissertation*. Oklahoma State University, Stillwater, OK.

Numrich R and Claus N. 1991. Druckabfall bel der kondensation in vertikalen rohren (in German). *Proceedings of Eurotherm Sem*. 18, 317-327.

Panchal CB and HuangFu EP. 2000. Effects of mitigating fouling on the energy efficiency of crude-oil distillation. *J Heat Transfer Engineering*. 21, 3-9.

Panchal CB, Kuru WC, Liao CF, Ebert WA, and Palen JW. 1997. Threshold conditions for crude-oil fouling. *Proceedings of the Engineering Foundation Conference: Understanding Heat Exchanger Fouling and Its Mitigation*, Begell House Publication, New York, NY. 273-281.

Kuru WC and Panchal CB. 1997. Application of single-phase fouling data to two-phase flows. *Proceedings of the Engineering Foundation Conference: Understanding Heat Exchanger Fouling and Its Mitigation*, Begell House Publication, New York, NY. 411-416.

Shah MM. 1976. A new correlation for heat transfer during boiling through pipes. *Trans. AHRAE 82*. 66-86.

Spedding PL, Woods GS, Raghunathan RS, and Watterson. 1998. Vertical two-phase flow - Part I: Flow regimes; Part II: Experimental semi-annular flow and hold-up; Part III: Pressure drop. *Trans Inst of Chem Eng*. 76, 620-634.

Spedding PL and Hand NP. 1995. Prediction of holdup and friction pressure loss in two-phase horizontal and downward co-current flow. *Proceedings of Int Symp of Two-Phase Flow Modeling* Experiment. Rome, Italy. 573-582.

Spedding PL and Nguyen VT. 1980. Regime maps for air water two-phase flow. *Chem Eng Sci*. 35, 779-793.

Steiner D and Taborek J. 1992. Flow boiling heat transfer in vertical tubes correlated by an asymptotic model. *Heat Transfer Eng*. 13, 43-69.

Taitel Y and Dukler AE. 1976. A model for predicting flow regime transitions in horizontal and near horizontal gas-liquid flow. *AIChE Journal*. 22, 47-54.

Wadekar VV and Kenning DBR. 1990. Flow boiling heat transfer in vertical slug and churn flow region. *Proceedings of Int. Heat Trans Conf.* Jerusalem. Israel. 449-454.

29a Pem Fuel Cells - Challenge in Science and Technology

Virendra K. Mathur, Dept. of Chem. Eng., University of New Hampshire, 260 Kingsbury Hall, Durham, NH 03824 and Xiaofeng Xie, Institute of Nuclear and New Energy Technology,Tsinghua University, Qinghua Yuan, Beijing, 100084, China.

Fuel cells are moving past the development stage and into trials for real applications. They are the most promising energy source for transportation, heating and power generation. In a fuel cell, hydrogen is ionized to produce an electrical current and then the hydrogen ion reacts with oxygen to form water and heat. The perceived benefits include high efficiency, efficiency relatively independent of load, reduced pollution, high power density, no moving parts, very low noise output, short start-up time and capable of automatic operation. The immediate difficulties to overcome include: difficult fuel storage, sensitivity to fuel impurities, maintaining proper moisture control of the membrane, market penetration and high cost to produce. There are a variety of fuel cells in different stages of development and they can be classified by the ion transportation separator that is used to facilitate the electrochemical reactions. The most common classification of fuel cells is by the type of electrolyte in the cells and includes 1) proton exchange membrane (polymer) electrolyte fuel cell (PEMFC); 2) alkaline fuel cell (AFC); 3) phosphoric acid fuel cell (PAFC); 4) molten carbonate fuel cell (MCFC) and solid oxide fuel cell (SOFC). PEMFC will be the fuel cell type primarily discussed in the chapter. PEMFC produce electricity via a direct electrochemical energy conversion process rather than a combustion process typical of the current fossil fuel power plants. These fuel cells operate at low temperature allowing for faster start-up and immediate response to changes in demands for power. The PEMFC system is seen as the system of choice for vehicular power applications, small stationary power applications and as battery replacements in small, hand-held devices. These fuel cells typically operate on hydrogen or methanol as fuel. This paper discusses the challenges in science and technology facing this industry.

Effects of Sub-Freezing Temperatures on A PEM Fuel Cell

Qiangu Yan[1,*], Hossein Toghiani[2], Young-Whan Lee[1], Kaiwen Liang[2]
[1] Center for Advanced Vehicular Systems (CAVS), Engineering Research Center (ERC), Box 5405, Mississippi State University, MS 39762-5405, Tel: (662)-325-5906, Fax: (662)-325-5433
[2] Dave C. Swalm School of Chemical Engineering, Box 9595, Mississippi State University, MS 39762

Abstract The cold-start behavior and the effect of subzero temperatures on fuel cell performance were studied using a 25-cm^2 PEMFC. The fuel cell system was housed in an environmental chamber that allowed the system to be subjected to temperatures ranging from sub-freezing to those encountered during normal operation. Fuel cell cold-start was investigated under a wide range of operating conditions. The cold-start measurements showed that the cell was capable of starting operation at -5 °C without irreversible performance loss when the cell was initially dry. The fuel cell was also able to operate at low environmental temperatures, down to –15 °C. However, irreversible performance losses were found if the cell cathode temperature fell below –5 °C during operation. Freezing of the water generated by fuel cell operation damaged fuel cell internal components. Several low temperature failure cases were investigated in PEM fuel cells that underwent sub-zero start and operation from –20 °C. Cell components were removed from the fuel cells and analyzed with scanning electron microscopy (SEM). Significant damage to the MEA and backing layer was observed in these components after operation below –5 °C. Catalyst layer delamination from both the membrane and the gas diffusion layer (GDL) was observed, as were cracks in the membrane, leading to hydrogen crossover. The membrane surface became rough and cracked and pinhole formation was observed in the membrane after operation at sub-zero temperatures. Some minor damage was observed to the backing layer coating Teflon and binder structure due to ice formation during operation.

INTRODUCTION

Proton exchange membrane fuel cells (PEMFCs) have attracted great attention in recent years as a promising replacement for traditional engines due to their high power density and ultra-low emission features [1-3]. Fuel cells also offer an opportunity to use an almost limitless fuel: hydrogen. Performance with increased stability and reliability as well as low cost must be realized before fuel cells can replace internal combustion engines [4]. PEM fuel cells usually operate at temperatures between 70 to 80 °C. However, in automotive applications, PEM fuel cells will be exposed to a wider range of temperature, including subzero temperatures. Therefore, PEMFCs must be able to operate under these conditions that might be found in colder climates. One potential problem is that water present in the fuel cell could freeze when the fuel cell temperature drops below 0 °C. Since power generation efficiency is low at temperatures below the normal operating temperature, starting the fuel cell from low temperature is quite difficult. Moreover, prolonged exposure of the fuel cell to sub-freezing temperatures may result in ice formation. This could potentially damage the internal structure and components of the fuel cell. During fuel cell cold start and sub-freezing operation, the water generated initially at the cathode will freeze. This can block the pathways through the porous gas diffusion layer and catalyst layer, thereby, preventing the movement of reactants. Subzero-freezing operation of fuel cell systems is an important concern; however, limited research efforts have focused on this area. The effect of water freezing on the PEMFC performance and its effect on PEMFC components have been examined by theoretical models [5-8], experimental methods [9-16], and fuel cell design optimization efforts [17-19]. Seshadri et al. [5] developed both steady state and transient models of a PEM fuel cell power plant for transportation applications. The results of the power plant performance testing indicated good agreement with these models. Testing showed that the power plant achieved stable performance at all power levels, including low power holds. Gummalla et al. [6] reported a study of bootstrap start using physics-based dynamic models. These models included a detailed fuel cell model, a semi-detailed stack model, and a system-level stack model. The detailed model captured the strongly coupled reaction and transport through the various sub-layers in a PEM fuel cell, namely the gas diffusion layers, catalyst layers and the polymer membrane. Comparison of model predictions and experimental data for voltage response during bootstrap start showed reasonable agreement, validating the

* Corresponding authors e-mail: yan@cavs.msstate.edu

modeling approach. The model was then deployed to analyze the startup behavior from sub-freezing conditions. The model predictions of start-up temperature dynamics and the voltage response were in good agreement with the test data. These models were used for conceptual and detailed design of fuel cell stacks and power plant system. The research of Sundaresan and Moore [7] focused on the development and use of a sub-freezing thermal model for a polymer electrolyte fuel cell stack. Specifically, the work focused on the generation of a model in which the fuel cell was separated into layers to determine an accurate temperature distribution within the stack. The layered model revealed the effect of the endplate thermal mass on the end cells, and accommodated the evaluation of internal heating methods that may mitigate this effect. To investigate the basic cold start behavior under isothermal conditions, Oszcipok et al. [8] used statistical analysis to study operational influences on the cold start behavior of PEM fuel cells. Their regression showed that the initial current mainly depended on the membrane humidity and the operating voltage. Cho et al. [9,10] investigated the effects of freezing/thermal cycles from 80 to −10 °C on PEMFC performance. They found that the cell performance was degraded due to the phase transformation and volume changes of water during thermal cycling. The effect of water removal on PEMFC performance degradation was also examined under freezing/thermal cycles [10]. The performance degradation rate was reduced to 0.06%, or -0.47%, if water was removed from the PEMFC with dry gas or with an antifreeze solution, respectively. Cargnelli et al. [11] operated a 50 W PEMFC system in a climatic chamber at -42 °C. For startup, they used a catalytic burner to heat up and humidify the reaction air. Chu et al. [12] operated a 50 W fuel cell system at -10 °C for 9 hr. Datta and Velayutham reported results of a 500 W fuel cell system which was operated without any problems in the Antarctic [13]. Kagami et al. [14-16] performed non-isothermal, galvanostatic single fuel cell experiments below 0 °C. At constant gas flow, they reported that the cathode electrode surface area was reduced by ice formation, and as a result, the reaction was inhibited. With the empirical assumption that the cathode surface area was diminished in proportion to the amount of frozen water, simulations were performed which described the voltage decay over time and correlated well with experimental results. Beneficial for cold starts are low current densities which prevent voltage decay by freezing of product water. There are over two hundred patents concerning fuel cell start up and operation at subzero temperatures. Some of these focus on improving fuel cell subzero start up and operation by fuel cell design optimization [17-19]. Though significant progress was achieved during recent years, more work is required in these areas including modeling, experimental measurements and fuel cell design optimization based on a judicious selection of available approaches [20].

In this work, experimental studies were directed towards identifying the impact of operating variables through changes in fuel cell performance and by examination of fuel cell components after loss of performance had been established. In these initial experiments, a 25 cm^2 fuel cell was employed. The fuel cell system was housed in an environmental chamber that allowed the system to be subjected to temperatures ranging from sub-freezing conditions to those encountered during normal operation. Care was taken to ensure that the entering feed gas streams had sufficient residence time to thermally equilibrate at the temperature of the environmental chamber. During operation at sub-freezing temperatures, humidification was not employed. Cell voltage measurements under load as a function of time from different sub-freezing starting conditions provided information regarding the minimum current density that could be realized without significant degradation in cell voltage. Changes in surface morphology, and MEA structure were analyzed by microscope analysis.

EXPERIMENTAL

2.1Preparation of MEA and assembly of single cell

The catalyst ink for the electrodes was prepared by mixing the catalyst powders (20 wt % Pt/C, E-Tek), Nafion solution, and isopropyl alcohol. Then the prepared catalyst ink was sprayed on the wet-proofed carbon paper and carbon cloth with platinum loading of 0.4 mg/cm^2 for both anode and cathode. The membrane electrode assembly (MEA) was fabricated by placing the electrodes at both sides of the pretreated Nafion 112, 115 and 117 membranes, followed by hot pressing at 140°C and 200 kg/cm^2 for 4 min. The active electrode area was 25 cm^2. A 25 cm^2 single cell was constructed from the prepared MEA, Teflon gasket, and flow-field plates on both sides of the MEA and then placed in an environment chamber to control the cell temperature from 80 to −15°C.

2.2 Performance of fuel cell under subzero temperature

The effect of subzero temperatures on the cell performance was studied by measuring the polarization curve of the cell. The single cell was first operated at room temperature; then the operation was going on;

and after the cell performance stabilized the temperature of the chamber was cooled to the desired temperature during fuel cell running. The polarization curves of the fuel cell and cathode temperature were recorded during operation. The polarization curve of the cell was measured in ambient laboratory conditions between each of the temperature chamber measurements. In order to reveal possible irreversible performance losses due to freezing of product water inside the cell, the polarization curves were compared with the curves measured before the measurement series. The polarization curves were measured by scanning the current from 0 to 750 mA/cm^2 with 20 mA steps and simultaneously measuring the cell voltage. The voltage was allowed to stabilize for 30 s at each measurement point before the next current step. Before each polarization curve measurement, the cell was let to stabilize at the average current density of 100 mA cm^{-2} for 5 min. If the comparison of polarization curves showed changes in the cell performance, the MEA and GDLs were replaced. Hydrogen and air gases were fed bypass external bubbler humidifiers, and stoichiometry of the fuel was 1.2 and stoichiometry of air was between 2 and 6. Operating pressure was 1 atm.

2.3 Sub-zero performance and sub-zero start-up

Cold-start measurements were performed at temperatures of −5, −10 and −15 °C. A 25 cm^2 single cell was first operated at room temperature or 80°C; and the operation stopped until the cell performance stabilized. To prevent performance degradation caused by water freezing, during lowering the cell temperature after operation, dry nitrogen or hydrogen gases was fed to the anode and dry air was feed to cathode at a gas flow rate of 300 mL/min until the cell voltage was decreased to zero. Then the environment chamber was cooled to desired subzero temperature, kept at that temperature for 1 h. After that, the hydrogen/Air gases were switched on and the cell was operated at the current density increasing from 0 to 100 mA cm^{-2} or 200 mA cm^{-2}.

2.4 Characterization of MEA

After exposure to subzero temperature cycling has led to a permanent performance loss of greater than 10%, the MEA was removed from the cell. The polymer electrolyte and the electrodes were separated to allow independent examination. These components will be compared with fresh components (membrane, GDL, catalyst, flow field plate). An assortment of characterization tools such as SEM/EDX, XPS, TEM, were used to characterize the MEAs. In particular, characterization was focused on identifying chemical and/or physical changes that have occurred because of the sub-zero operating.

RESULTS AND DISCUSSION

3.1. Effect of subzero temperatures on PEM fuel cell performance

The effect of subzero temperatures on the cell performance was studied by measuring the polarization curve of the cell. The single cell was first operated at room temperature; then the operation was going on; and after the cell performance stabilized the temperature of the chamber was cooled to the desired temperature during fuel cell running. The chamber temperature ($T_{chamber}$) stabilized to the desired value in about 1 h except $T_{chamber}$ = -15 °C. The polarization curves were measured by scanning the current from 0 to 750 mA/cm^2 with 20 mA steps and simultaneously measuring the cell voltage. The voltage was allowed to stabilize for 30 s at each measurement point before the next current step. The polarization curves of the fuel cell and cathode temperature were recorded during operation. The polarization curves at different chamber temperatures are illustrated in Figure 1, and measured cell cathode temperatures are plotted in Figure 2.

Figure 1 shows fuel cell voltage at different chamber temperatures between -15 and 80^{0}C with the hydrogen stoic of 1.2 and air stoic of 2 and both feed gases without humidified. The polarization curves of the cell at different environmental temperatures show that the fuel cell performance is improved with increasing temperatures from -15 to 25 ^{0}C. The fuel cell has the best performance at room temperature (25 ^{0}C). However, the cell performance decreases as the temperature increase further to 80^{0}C. This observation is believed to result from the balance of increase in gas diffusivity and decrease in membrane conductivity at higher temperature. Water is easy to condense at lower temperature; water flooding may block the gas transport in catalyst layer and gas diffusion layer (GDL). Gas diffusivity of the fuel cell is improved with increasing cell temperature, however, membrane conductivity will decrease at high temperature since gas relative humidity and water content in membrane decrease. As a result, the cell performance is worse when temperature rises to 80 ^{0}C under operation conditions.

Fuel cell performance at chamber temperature of 80 ºC, 25 ºC, 0 ºC, -5 ºC and -10 ºC was stable and repeatable. However, when the chamber temperature was down to at -15 ºC, fuel cell performance was unstable, after the 4th current-voltage (CV) cycles the cell voltage was decreased and suddenly collapsed to zero at current density of 350 mA/cm^2 when the cell cathode temperature was below –5 ºC.

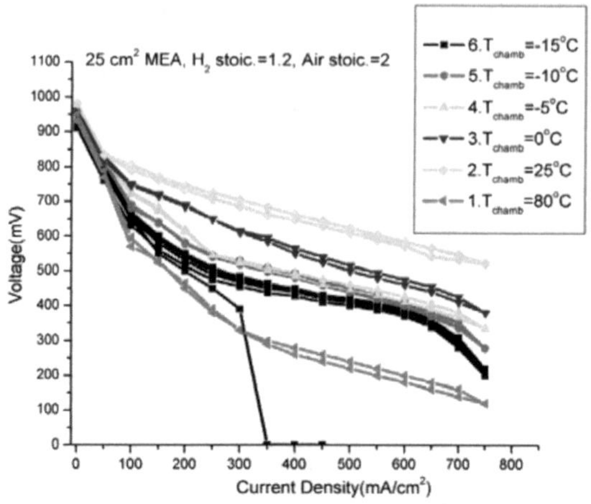

Figure 1. Polarization curves of a 25 cm^2 fuel cell under different subzero temperature.

The properties of Nafion at freezing temperatures can be quite relevant, for example, within the context of fuel cells in vehicles with regard to cold-starting, as well as the degradation of membrane/electrode assemblies due to the freezing of in situ water [21]. The state of water in Nafion was classified into two categories [22-24], nonfreezing, and freezing bound, and free water (bulk water) while there are three states for water over membrane surface, nonfreezing, freezing bound, and free water (bulk water). Free water does not have chemical interacting with ion within membrane which will freeze at around 0 °C; By means of differential scanning calorimetry and IR spectroscopy, Okada et al. [25] has demonstrated there are two types of water in the Nafion membrane: one type that does not freeze, and one type that freezes. (a) freezing water molecules, which interact with the cations and ion exchange sites weakly and could freeze at around -20 °C , behaving in a manner close to but not identical to that of bulk water and (b) nonfreezing water molecules which could not be solidified to ice even at -120 °C, and are considered as both the one binding strongly with the cations and the ion exchange sites and the ones exposed to fluorocarbon environment. Therefore, water inside Nafion membrane does not freeze at the temperature above –20 °C, as a result bulk Nafion membrane should not be affected and damaged at the temperature above –20 °C. However, if free water exists over membrane surface, it may freeze and damage the membrane. Free water usually freezes at around 0 °C, however, it may shift to lower temperature with decreasing water content in membrane or it is trapped in porous materials. There are also several states for water confined in porous materials. Hayashi et al. [26] classified water absorbed in porous coal into three types with different properties; free water that freezes around 273 K as normal bulk water, bound water that freezes at 213-260 K, and nonfreezable water. The water freezing temperature is affected by the pore size which water is filled and material properties. Bound water and nonfreezable water should present in pores which diameter is smaller than 10 nm based the Gibbs-Thompson equation [26]. The authors also analyzed the distribution of freezing point temperature of bound water in some coals on the basis of the Gibbs-Thompson equation. Based on the fact that the electric permittivity of water will decrease significantly after it is frozen [27] (Szurkowski *et al.*, 1993), Sliwinska-Bartkowiak et al. [28] used dielectric relaxation spectroscopy to determine the melting point of water in active carbon fiber with mean pore diameter around 1.8nm, and the capacitance of the system was found to change significantly with water phase transition at 273 K (0 °C) (bulk freezing temperature) and 235 K (-38 °C) (corresponding to complete water crystallization inside pores). The pore size distribution of catalyst layer and gas diffusion material carbon paper/colth ranges from several hundred micro-meters (μm) to several nanometers (nm), it significantly affects the water freeze-thaw process in MEAs. A portion of water will freeze if fuel cell is running under sub-zero temperature since water is generated at cathode side by oxygen reduction reaction (ORR) process, bulk water must exist in porous catalyst layer and GDL at sub-zero temperature. Therefore the MEA (include catalyst, GDL and membrane) would undergo damage after freezing, especially to the micro-pore structures. These micro-pore media could contain significant water in the cathode side when fuel cell running. Water trapped within the MEA may freeze with operation from sub-zero temperatures forming ice crystals within a portion of the pores. Fuel cell component like MEA structural degradation may occur as a result.

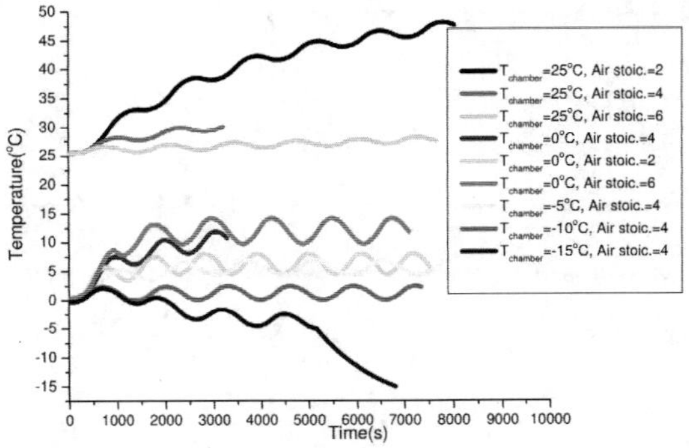

Figure 2. Cell cathode temperature of the fuel cell versus time during running under different environmental temperature.

The cell temperatures measured from the cathode side when the polarization curves were measured by scanning the current from 0 to 750 mA/cm^2 with 20 mA steps, and the results are illustrated in Figure 2. The cell temperatures are fluctuating when scanning the current. The cell temperatures were waved between 3 to 4 °C. The cell temperature increased simultaneously with the current scanning and reached a stable value. Fuel cell cathode temperature at chamber temperature of 25 ^{O}C, 0 ^{O}C, -5 ^{O}C and -10 ^{O}C was stable and repeatedly. However, chamber temperature at -15 ^{O}C, cathode temperature was unstable, after 4 cycles, the cell cathode temperature was below –5 ^{O}C, the cell performance was lost irreversible and cathode temperature decrease sharply to chamber temperature: –15 ^{O}C.

3.2. Cold-start measurements

Cold-start measurements were performed at chamber temperatures of −5, −10 and −15°C. A number of tests were conduced to evaluate the effects of operation conditions on a fuel cell startup. A 25 cm^2 single cell was first operated at room temperature; and the operation stopped until the cell performance stabilized. When normal operation of the fuel cell was ended, the fuel cell was pretreated with different methods before freezing to subzero temperature. Some of the fuel cells were purged with dry reactants, i.e., hydrogen and air to anode and cathode receptively. Some of the fuel cells were purged with dry nitrogen and air. Others were not. The fuel cells were then cooled to sub-freezing temperatures that varied from about 0°C to about −15°C, and allowed to stabilize for one hour. To restart the cells after freezing, the reactants (i.e. air and H_2) were supplied at room temperature without humidification. The Effects of chamber temperature, insulation, gas purging, air stoichiometry, and feed gas temperature on fuel cell startup are all investigated. Performance (cell voltage), cell temperature, and success or failure, were achieved. In these tests, the current was drawn from the cell (i.e. the load) at 100mA/cm^2 or 200 mA/cm^2. The results of those tests are shown in Table 1.

Table 1 Cold Star-up Cases*

$T_{chamber}$	N_2 purge	Insulate	Feed Gas Preheated	Air Stoic	Result
-5 °C	No	No	No (R.T.)	2	Failure
-5 °C	Yes	No	No (R.T.)	2	Failure
-5 °C	Yes	Yes	No (R.T.)	2	Success
-10 °C	Yes	Yes	No (R.T.)	2	Failure
-10 °C	Yes	Yes	No (R.T.)	4	Success
-15 °C	Yes	Yes	No (R.T.)	4	Failure
-5 °C	Yes	No	Yes (60 °C)	2	Success
-10 °C	Yes	No	Yes (60 °C)	2	Failure
-10 °C	Yes	No	Yes (80 °C)	2	Success

*A 25 cm^2 Fuel cell, H_2 stoic. =1.2, all the feed gases were dry, current density stepped from 0 to 100 mA/cm^2 when start up.

The cold-start behavior of the cell at −5 °C is illustrated in Figure 3 and at −10 and −15 °C in Figure 4. The cell was capable of starting operation at −5 °C without any difficulties if the cell was pre-purged and insulated. If the fuel cell was shut down without purge, therefore some free water present on the catalyst and in the gas diffusion layers of the cell, free water may freeze to ice when the fuel cell is subjected to freezing temperatures, hence block the penetration of reactants into the catalyst from the gas diffusion layer and prevent the desired chemical reaction from occurring. In this failure case, fuel cell cannot startup as the result in table 1 and Figure 3. If the cell was purged using dry H_2 and air before shutdown, the cell was then decreased to 5 °C, the cell was startup (a load was connected with current density of 100 mA/cm^2 or 200 mA/cm^2) in the after approximately 20 seconds while the voltage decreased but recovered gradually, the fuel cell temperature increased gradually (Figure 4). Once operation of the fuel cell had startup, the exothermic reaction of hydrogen and oxygen within the fuel cell and the resistive heating due to internal ohmic losses caused the cell temperature to rise, the feed gases heated the cell is not neglected either, therefore, the preheat of feed gas also affected the fuel cell startup from subzero temperature according to Table 1, Figure 4 and Figure 6 results. Air can carry away the water formed in the cathode catalyst layer before it accumulates and frozen in the cell, therefore increasing air flow rate will result in improvement fuel cell operation under subzero temperature. .

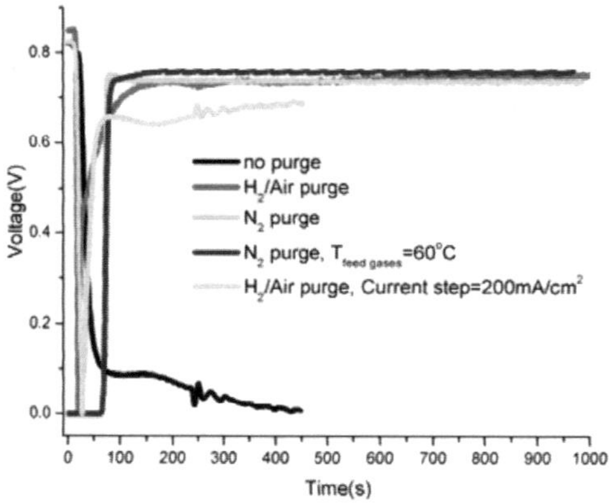

Figure 3 The plots of fuel cell voltage versus time after startup at –5 °C.

The cell was capable of starting operation with the proper starting procedure at −10 °C (Figure 5). With the higher air stoichiometry and higher feed gas temperaure at −10 °C the cell can be start up with the current step from 0 to 100 mA/cm^2, but the cell failed to start up if the step current increase to 200 mA/cm^2. In the case of lower air stoichiometry and feed gas temperature, the cell also failed to start up, this may due to the cell was not heated rapidly enough with the starting procedure to prevent fatal freezing of the product water inside the cell.

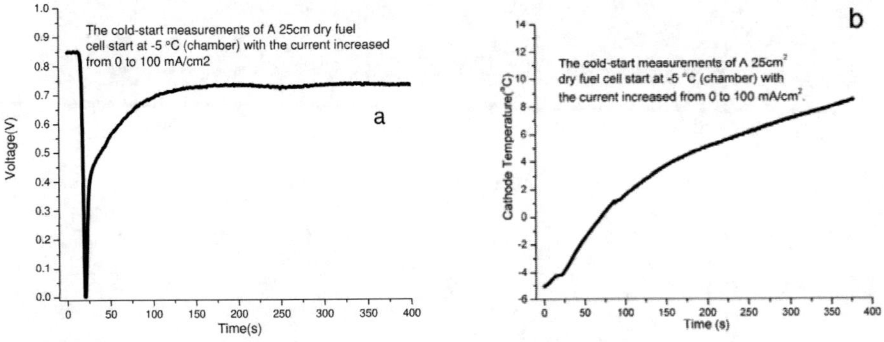

Figure 4 The plots of fuel cell voltage (a) and cell cathode temperature (b) versus time after startup at −5 °C.

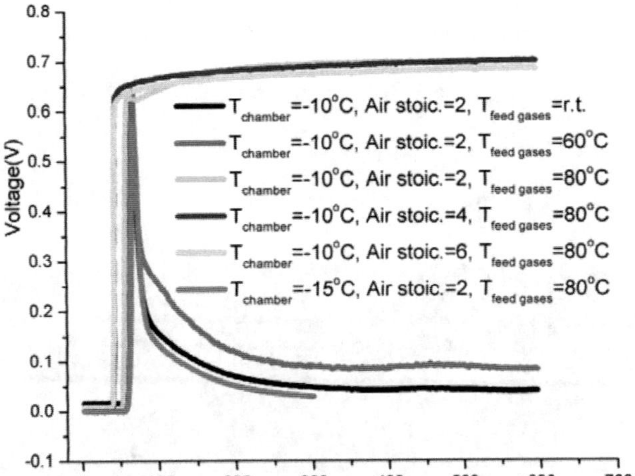

Figure 5 The plots of fuel c plot of fuel cell voltage versus time after startup under −10 and −15 °C.

3.3 Effects of freezing temperature on fuel cell components

Figures 6 to Figure 8 illustrate the difference in MEA, gas diffusion layer and membrane damage observed in fuel cell after operation failure or performance degradation. Images of new samples are provided for reference and to illustrate the overall magnitude of the damage.

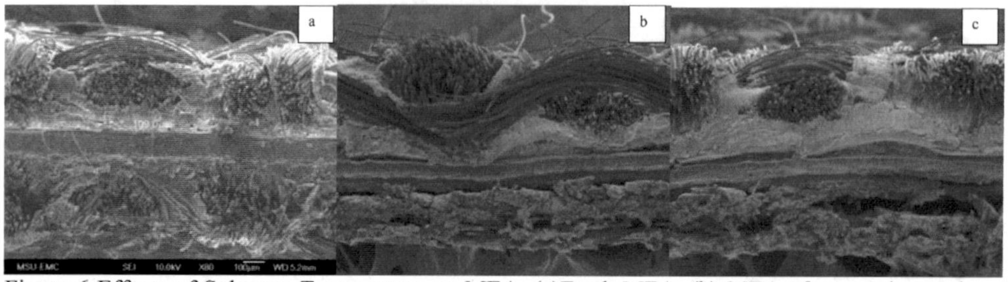

Figure 6 Effects of Sub-zero Temperature on MEA. (a)Fresh MEA, (b) MEA after running under –10 °C, (c) MEA after running under –15 °C.

Figures 6 illustrate the massive damage observed at an MEA taken from failure fuel cell after running at the chamber temperature of –15 °C. Freezing of water has delaminated the catalyst layer off both the membrane (Figure 6 b) and GDL (Figure 6 c). Figure 7 shows that some damage to the backing layer fibers and binder is occurring to the failure fuel cell. Figure 7c examines damage at a smaller scale, from this figure; it does appear that damage is occurring to the backing layer binder structure on a very small scale. It is apparent from Figure 7 that GDL which trap or hold bulk water during operation undergo damage when exposed to freezing temperatures. Some minor damage to the backing layer fibers and binder is occurring when GDL exposed to freezing temperatures. Figures 8 shows the damage to the membrane. In the low magnification image, damage is present in the sample of the failure fuel cell since the membrane surface became rough and crush after it underwent subzero freezing (Figure 8b). In the high magnification image (Figure 8c,d), the membrane does appear to show signs of pinhole (Figure 8c) and micro cracking (Figure 8d) following the freezing operation.

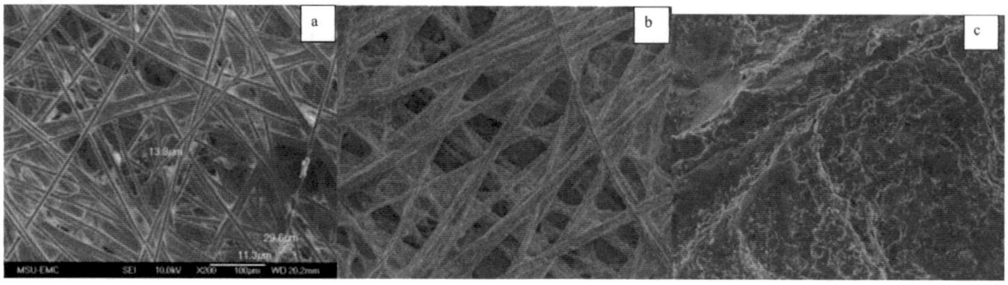

Figure 7 Effects of Sub-zero Temperature on GDL. (a) Fresh carbon paper, (b) Carbon paper after running under 15 °C, (c) Higher magnification image of carbon paper after running under 15 °C.

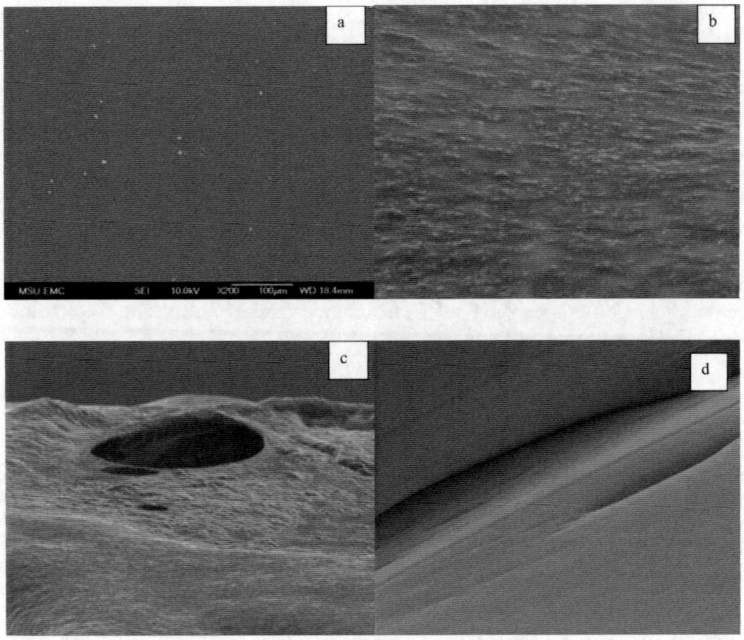

Figure 8 Effects of Sub-zero Temperature on Membrane. (a) Fresh Nafion membrane, (b)Membrane after running under -15 °C, (b) Membrane after running under -15 °C, (c) Membrane from cathode outlet regions after running under -15 °C, (d) Membrane from cathode outlet regions after running under -15 °C.

CONCLUSIONS

The effect of subzero temperatures on the performance of a PEMFC was studied. Polarization curves were measured at chamber temperatures between 80 and −15 °C. The cell cathode temperatures were also recorded when each polarization curve was measured. Fuel cell performance at chamber temperatures of 25 °C, 0 °C, −5 °C and −10 °C was stable and reproducible. The cell temperature fluctuated as the current was scanned, with fluctuations between 2 to 3 °C. The cell temperature, however, achieved a stable shape. The fuel cell cathode temperature at chamber temperatures of 25 °C, 0 °C, −5 °C and −10 °C was stable and reproducible. However, at a chamber temperature of −15 °C, fuel cell performance was unstable. After four cycles, the cell voltage was decreased and suddenly collapsed to zero at a current density of 350 mA/cm^2. This sudden collapse occurred with the cathode temperature below –5 °C. . The cathode temperature was unstable and suddenly dropped sharply to −15 °C.

Cell cold-start was investigated at temperatures of −5 °C, −10 °C and −15 °C. Results demonstrated that a single cell was able to start at −5 °C if the cell was pre-purged and insulated. If the fuel cell was shut down without purge, it could not be subsequently started up. The cell was capable of starting operation with the proper starting procedure at −10 °C. With the higher air stoichiometry and higher feed gas temperature at −10 °C, it was possible to start up the cell using a current step from 0 to 100 mA/cm^2. However, if the current step was 0 to 200 mA/cm^2, the cell failed to start up. SEM results illustrated that the MEA, gas diffusion layer and membrane in the fuel cell were damaged leading to system failure or performance degradation. Delamination of the catalyst layer from both the membrane and GDL was found have occurred in the MEAs exposed to subzero temperatures during operation. Minor damage to the backing layer Teflon coating was also found.

ACKNOWLEDGMENTS

This project is generously supported by the Center of Advanced Vehicular Systems (CAVS) at Mississippi State University.

REFERENCES

1. D Oliver J. Murphy, G. Duncan Hitchens, David J. Manko, J. of Power Sources, 47 (1994) 353-368.
2. S. J. C. Cleghorn, X. Ren, T. E. Springer, M. S. Wilson, C. Zawodzinski, T. A. Zawodzinski, S. Gottesfeld, International Journal of Hydrogen Energy, 22 (1997) 1137-1144
3. Keith B. Prater, J. of Power Sources, 61 (1996) 105-109
4. Jian Xie, David L. Wood, III, Karren L. More, Plamen Atanassov, Rodney L. Borup, J. Electrochem. Soc. 152, (2005) A1011
5. Parthasarathy Seshadri, Zakiul Kabir, Editor(s): Ramesh K.Shah, Etim U.Ubong, Scott Samuelsen, 3rd International Conference on Fuel Cell Science, Engineering, and Technology, Proceedings, Ypsilanti, MI, United States, May 23-25, 2005 (2005), 733-738
6. Mallika Gummalla, Nikunj Gupta, Shubhro Ghosh, Sergei Burlatsky, Patrick Hagans, Cynthia Rice, AIChE Annual Meeting, Conference Proceedings, Austin, TX, United States, Nov. 7-12, 2004 (2004), 023A/1-023A/7.
7. M. Sundaresan, R. M. Moore, Journal of Power Sources (2005), 145(2), 534-545
8. M. Oszcipok, D. Riemann, U. Kronenwett, M. Kreideweis, M. Zedda, Journal of Power Sources (2005), 145(2), 407-415.
9. E. A. Cho, J.-J. Ko, H. -Y. Ha, S.-A. Hong, K.-Y. Lee, T.-W. Lim, I.-H. Oh, J. Electrochem. Soc., 150, (2003) A1667
10. Eun Ae Cho, Jae-Joon Ko, Heung Yong Ha, Seong-Ahn Hong, Kwan-Young Lee, Tae-Won Lim, In-Hwan Oh, J. Electrochem. Soc. 151, (2004) A661
11. J. Cargnelli, P. Rivard, D. Frank, R.B. Gopal, Fuel cell for deployable applications in sub-zero environment, Portable Fuel Cells (1999).
12. D. Chu, R. Jiang, K. Gardner, R. Jacobs, J. Schmidt, T. Quakenbush, J. Stephens, Polymer electrolyte membrane fuel cells for communication applications, J. Power Sources 96 (2001).
13. B.K. Datta, G. Velayutham, A. Prasad Goud, Fuel cell power source for a cold region, J. Power Sources 106 (2002).
14. F. Kagami, T. Ogawa; Y. Hishinuma, T. Chikahisa, Simulating the performance of a PEFC at a temperature below freezing, Fuel cell seminar, Abstracts, 2002.
15. F. Kagami, Y. Hishinuma, T. Chikahisa, Performance and selfstarting of a PEFC at temperatures below freezing, Thermal Sci. Eng. 10 (3) (2002).
16. F. Kagami, T. Ogawa, Y. Hishinuma, T. Ghikahisa, *2002 Fuel Cell Seminar*, p. 239 (2002)
17. Eric L. Thompson, Robert L. Fuss, United States Patent 6887598
18. Richard J. Assarabowski, William T. Unkert, Leonard A. Bach, Albert P. Grasso, Benoit Charles Olsommer, United States Patent 6797421
19. Andreas Christen, Jens Thomas Mueller, United States Patent 6743538
20. Jean St.-Pierre, Joy Roberts, Kevin Colbow, Stephen Campbell, Amy Nelson, Journal of New Materials for Electrochemical Systems (2005), 8(3), 163-176.
21. A. Mauritz, R. B. Moore, State of Understanding of Nafion, Chem. Rev., 2004, 104(10), 4535-4586
22. Yoshida, H.; Miura, Y. *J. Membr. Sci.* 1992, *68*, 1.
23. Pineri, M.; Volino, F.; Escoubes, M. *J. Polym. Sci., Polym. Phys. Ed.* 1985, *23*, 2009.
24. N. Boyle, V. J. McBrierty, A. Eisenberg, *Macromolecules*, 1983, *16*, 80.
25. Gang Xie, Tatsuhiro Okada, Denki Kagakuoyobi Kogyo Butsuri Kagaku (1996), 64(6), 718-726
26. Hayashi, J., Norinaga, K., Kudo, N., and Chiba, T., Estimation of size shape of pores in moist coal utilizing sorbed water as a molecular probe, *Energy & Fuels* 15, 903-909, 2001.
27. B. Szurkowski, T. Hilczer, M. Sliwinska-Bartkowiak, *Ber. Bunsen-Ges. Phys. Chem.*, 97(5), 731, 1993.
28. M. Sliwinska-Bartkowiak, G. Dudziak, R. Sikorski, R. Gras, K. E. Gubbins, R. Radhakrishnan, Dielectric studies of freezing behavior in porous materials: Water and methanol in activated carbon fibers, *Phys. Chem. Chem. Phys.* 3, 1179-1184, 2001.

29c Electrochemical Impedance Spectroscopic Studies of Non-Fluorinate Proton Exchange Membranes in Dmfc

Xiaofeng Xie[1], Yuming Shang[1], Jiangwei Guo[1], and Virendra K. Mathur[2]. (1) Institute of Nuclear and New Energy Technology,Tsinghua University, Qinghua Yuan, Beijing, 100084, China, (2) Dept. of Chem. Eng., University of New Hampshire, 260 Kingsbury Hall, Durham, NH 03824

In direct methanol fuel cell (DMFC), protons and carbon dioxide are catalytically produced at the anode. The protons migrate through the membrane and react catalytically with oxygen to produce water at the cathode. Methanol transported across the membrane can react at the cathode without producing electricity. This not only lowers fuel utilization, but also adversely effects cathode performance. A successful DMFC membrane must not only conduct protons, but also prevent methanol transport.

This work explores the potential of novel membranes in DMFC. Hybrid membranes of inorganic and organic materials are developed in our studies. The membranes have demonstrated characteristics that suggest about 10-fold improvement in DMFC. This improvement may be due to an alternate form of proton conduction in which protons travel via a Grotthus or ``hopping" mechanism. The characteristics of a membrane and its effects on the performance of fuel cell are investigated by means of electrochemical impedance spectroscopy (EIS). The EIS results have shown that ohm resistance, charge transfer resistance and mass transport resistance increase with discharging time. At high current densities, the water content on the anode side of the electrolyte increases due to electro-osmotic drag. As a consequence, the performance loss is probably due to water shortage at the outlet of cathode obstructing our passive DMFC to breathe air.

Key words: direct methanol fuel cell, electrochemical impedance spectroscopy, membrane electrode assembly

Photoproduction of Hydrogen from Hydrogen Sulfide using a Novel Metal Oxide Photocatalyst

Jin-Ook Beag*, ***Sang-Jin Moon, Chul Wee Lee, and Hyunju Chang***

Advanced Chemical Technology Division, Korea Research Institute of Chemical Technology, 100 Jang-Dong, Yusong, Daejon 305-600, Republic of Korea

Abstract

The decomposion of hydrogen sulfide has attained a lot of significance due to environmental problem. The Claus process is the most popular process that currently used for the hydrogen sulfide decomposition. The conventional Claus process is involved the synproportionation reaction between H_2S and SO_2 yielding elemental sulfur and water vapor. Two hydrogens in H_2S are wasted to make H_2O. If we could recover hydrogen from H_2S, instead of oxidizing it into H_2O (Claus process), immense amount of hydrogen could be produced. Photocleavage of the H_2S to H_2 by semiconductor particle photocatalyst has been considered as a process that ultimately converts solar energy into fuels. This process requires less energy compared to photodecomposition of water. In this connection, recovery of H_2 from H_2S has attracted more attention because of its economic and environmental advantages. Of late, development of photocatalyst for photodecomposition of hydrogen sulfide and water into hydrogen has been a subject matter of immense interest owing to the energy and pollution abatement. Massive work has been carried out in the development of ultraviolet driven photocatalysts for water and H_2S splitting. However, there is a demand for highly efficient photocatalyst for photoproduction of hydrogen. It is indispensable to control the energy structure for the development of photocatalyst materials. So, there is a need to find stable, economical and efficient new metal oxide photocatalysts for hydrogen production.

In view of this, we have investigated novel photocatlysts for photodecomposition of H_2S for the first time. We investigated the synthesis of this catalyst by solid state and solution route. The structural and morphological study was carried out using XRD and FESEM. The influence of preparation method on the activity of the photocatalyst was investigated. The novelty in synthesis of new product, its characterization and photocatalytic activity will be disscussed

Key words : photocatalyst, hydrogen, photodecomposition, metal oxide

1. Introduction

Among various method of solar energy conversion, much attention has been paid to photocatalytic decomposition of H_2S splitting for its potential significance in obtaining directly clean and high energy containing H_2 from abundant H2S. Every year, millions tons of H_2S are produced in the world with natural gas or from refinery plants, it will not remain constant but it expected to increase in the future. Moreover, hydrogen is highly valued as an energy source and an important chemical feedstock for organic synthesis. The direct thermal decomposition of H2S is thermodynamically unfavorable up to 1800 K. Photochemical decomposition is another alternative method for H_2S splitting, but the efficiency of this method depends on the activity of photocatalyst. If successfully developed with an economic viability, this could be the ultimate technology that could solve both energy and environmental problems altogether in future.

Therefore, development of an active photocatalyst under visible light region has attained immense importance and has become an imperative topic in photocatalysis research today. Doping of foreign transition elements into an active existing photocatalysts was studied extensively [1-4]. Especially, Ni and Cu-doped ZnS photocatalysts show activity for H2 evolution under visible-light irradiation even without Pt catalyst. Many researchers have reported the doping method for CdS, ZnS, VS, VS_4, WO_3 [5-8]. The nitrides (Ta_3N_5), oxynitrides (TaON, $LaTiO_2N$) and oxysulfides ($Sm_2Ti_2S_2O_5$) have been reported as active photocatalysts [9-13].

Another popular route to synthesis a visible light photocatalyst is to replace oxygen by N, C, S etc into the large band gap noble metal oxide anionic network. Domen et al showed that Ta–O–N system energy gap could be tuned by adjusting the amount of nitrogen [14] and therefore, the electrical properties of the material can be modified.

In this paper, we report the synthesis of niobium zirconium oxynitride ($Nb_2Zr_6O_{17-x}N_x$) from $Nb_2Zr_6O_{17}$ by ammonolysis method. The as-synthesized material in its pristine state was found to be an excellent photocatalyst for the decomposition of H_2S to hydrogen.

2.Experimental

2.1 Synthesis $Nb_2Zr_6O_{17}$ by solid-state reaction (SSR) route

All samples in this study were prepared from reagent-grade oxides: ZrO_2 (99.9%, Aldrich, USA) and Nb_2O_5 (99%, Strem Chemicals, USA). Appropriate amounts of raw materials for stoichiometric ($Nb_2Zr_6O_{17}$) were mixed and ground well then taken in alumna crucible to calcined at 1573 K for 24 h. After cooling to room temperature the calcined product was ground, washed with distilled water then it was dried at 373 K referred as $Nb_2Zr_6O_{17}$.

2.2. Syntheses of $Nb_2Zr_6O_{17-x}N_x$ by thermal ammonolysis route

$Nb_2Zr_6O_{17-x}N_x$ was prepared by heating $Nb_2Zr_6O_{17}$ as synthesized from the above procedure in a quartz tube reactor under a flow of ammonia gas (flow rate: 20 –100 cm^3 min^{-1}) at 1073 K for 2 h. After the reaction times, the heating furnace was switched off and the product powders were allowed to cool to room temperature under pure argon atmosphere, The rose red color powder was obtained is referred as $Nb_2Zr_6O_{17-x}N_x$.

2.3 Pt and RuO_2 doping

$Nb_2Zr_6O_{17-x}N_x$ catalyst was doped with 1 wt% of Pt and RuO_2 using $H_2PtCl_6$and $RuCl_3$ respectively. Thus, the catalyst was dispersed in water (methanol for Pt) and an appropriate quantity of dopant was added and the resultant suspension was sonicated for 30 minutes. The solvent was removed on a rotary evaporator and finally oxidized in air for 3 h at 673 K for Pt while for RuO_2 at 973 K. The Pt doped oxidized materials were reduced by H_2 (20 ml/min flow rate) for 2h at 773 K and reoxidized at 473 K for 1 hr. While RuO_2 doped oxidized materials were used as such.

2.4 Characterization

Powder X-ray Difractograms were recorded on a Model Rigaku-D/MaX-2200V X-ray Difractometer with CuK∝ radiation with Ni filter. The surface morphology and particle size were determined using a Field Emission Scanning Electron Microscope (FESEM Model JEOL-JSM6700F). Transmission Electron Microscopy (TEM) was performed using FE Tecani G^2 (200 KeV). X-ray photoelectron (XPS) spectra were recorded from the prepared samples with ESCA-LABMK-II-LTD analyzer under a vacuum better than 1×10^{-8} Torr, using MgK□radiation and a constant pass energy of 50 eV. The DRS of the as-synthesized photocatalysts was measured by using UV-visible spectrophotometer (Model SHIMADZU UV-2450 at diffuse reflectance mode). The baseline correction was done using calibrated sample of $BaSO_4$ between 200 to 800 nm.

2.5 Measurements of photocatalytic activity

Cylindrical Pyrex photochemical reactor with a quartz window and thermostat water jacket was used for the photocatalytic measurements. A Xe-lamp light source (Oriel) of intensity 450 W with cut-off filter (≥420 nm) was used for the undoped and doped samples. Each experiment was conducted by using 0.5 g of catalyst in 250 ml of KOH solution (0.5M) with H_2S flow, 2.5 ml/min. The vigorously stirred suspension was purged with argon for 1h followed by bubbling of hydrogen sulfide (H_2S) for about 1h at 25 ^oC. The excess hydrogen sulfide was trapped in NaOH solution. The amount of evolved hydrogen was measured using graduated gas burette and gas chromatograph (Model Shimadzu GC-14B, MS-5Å column, TCD, Ar gas carrier).

The apparent quantum yield [15] was measured using a 450W Xe lamp (Oriel) attached with an IR filter (Oriel), an interference filter (430 ± 5nm; Oriel) and a cutoff filter ($\lambda \geq 420$nm; Oriel). The number of incident photons were measured using a silicon photodiode with integrating sphere (Oriel). The apparent quantum yield (%) is calculated by the following equation.

$$\text{Quantum Yield (\%)} = \frac{[\text{Number of } H_2 \text{ molecules evolved x 2}]}{[\text{Number of incident photons}]} \times 100$$

3. Results and Discussion

3.1 XRD Analysis

Powder XRD patterns of $Nb_2Zr_6O_{17}$ and $Nb_2Zr_6O_{17-x}N_x$ samples are as shown in Fig.1. The XRD patterns (Fig. 1a) of $Nb_2Zr_6O_{17}$ showed all the peaks (d-values) are well matching with the reported data of orthorhombic $Nb_2Zr_6O_{17}$ (JCPDS card No. 72-1745), indicates the reaction was completed for to the proposed stoichiometry of precursors at desired conditions. The oxynitride samples ($Nb_2Zr_6O_{17-x}N_x$) (Fig. 1b) showed similar XRD patterns (orthorhombic $Nb_2Zr_6O_{17}$) with little shifting of 2θ values of each XRD peaks demonstrate that nitrogen might be doped in the material. Therefore, the oxynitride material may have the molecular formula $Nb_2Zr_6O_{17-x}N_x$. The detail crystal structure of these as-synthesized materials is also under investigation.

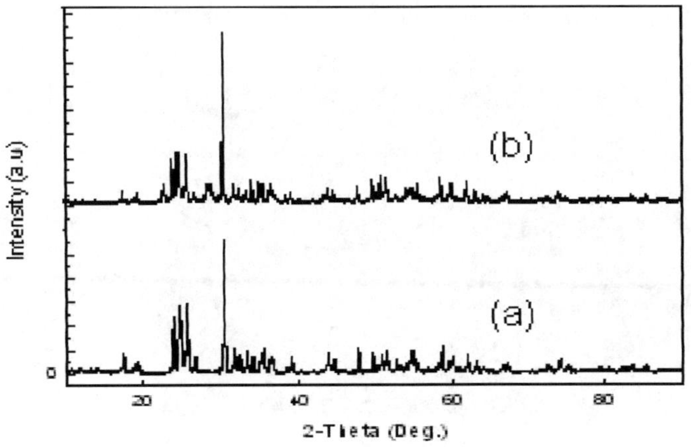

Fig. 1. XRD patterns of (a) $Nb_2Zr_6O_{17}$ and (b) $Nb_2Zr_6O_{17-x}N_x$ as synthesized samples

3.2 Electronic band structure by DRS

Ultraviolet-visible diffuse reflectance spectra (UV-DRS) of $Nb_2Zr_6O_{17}$ and $Nb_2Zr_6O_{17-x}N_x$ as synthesized materials are as shown in Fig. 2. The nature of absorption band of $Nb_2Zr_6O_{17}$ suggests that samples are homogeneous with pure phase indicating the completion of the reaction for the proposed stoichiometry but the absorption edge ≤ 400 nm (~ 3.1 eV) is not in visible range (Fig.2a). While the ammonolysis oxynitride sample showed a red shift (Fig.2b) as compared to its mother material (Fig. 2a). The absorption cut of the $Nb_2Zr_6O_{17-x}N_x$ showed the absorption in visible region at 580 nm (2.14 eV) (which is the major source of solar energy) indicates the nitrogen doped in the oxide samples.

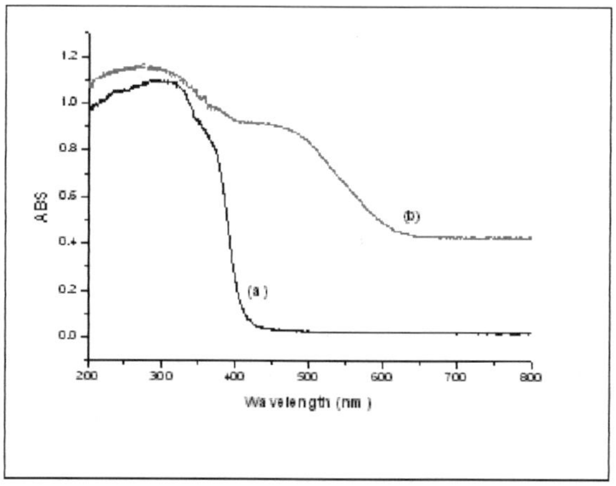

Fig. 2. DRS of (a) $Nb_2Zr_6O_{17}$ and (b) $Nb_2Zr_6O_{17-x}N_x$ as synthesized samples

3.3 Particle morphology of $Nb_2Zr_6O_{17}$ and $Nb_2Zr_6O_{17-x}N_x$ by SEM

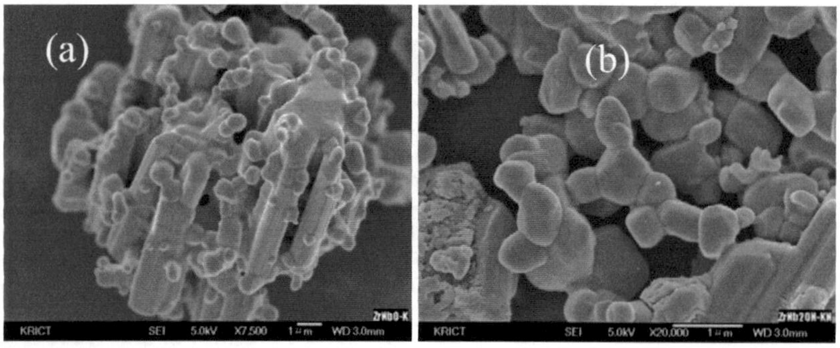

Fig. 3. SEM micrographs of (a) $Nb_2Zr_6O_{17}$ and (b) $Nb_2Zr_6O_{17-x}N_x$ as synthesized catalysts

Fig. 3 represents Field Emission Scanning Electron Microscopy (FESEM) images of $Nb_2Zr_6O_{17}$ and $Nb_2Zr_6O_{17-x}N_x$. The FESEM pictures of $Nb_2Zr_6O_{17}$ (Fig. 3a) showed agglomeration of uneven elongated orthorhombic particles. While the $Nb_2Zr_6O_{17-x}N_x$ (Fig. 3b) yielded prismatic pseudo orthorhombic shaped particles with some agglomerates and also less denser indicates more porous as compared to its mother material. These agglomerates were found despite the samples were sonicated prior to SEM analysis indicating strong Van der Waal's attraction between the nanoparticles. Three batches of the same material showed the reproducible shape and size, but with little randomization of homogeneous shapes. SEM photographs have been taken from randomly selected areas of the substrate and as such these are representative of the overall sizes.

3.4 Particle morphology of $Nb_2Zr_6O_{17-x}N_x$ *by TEM*

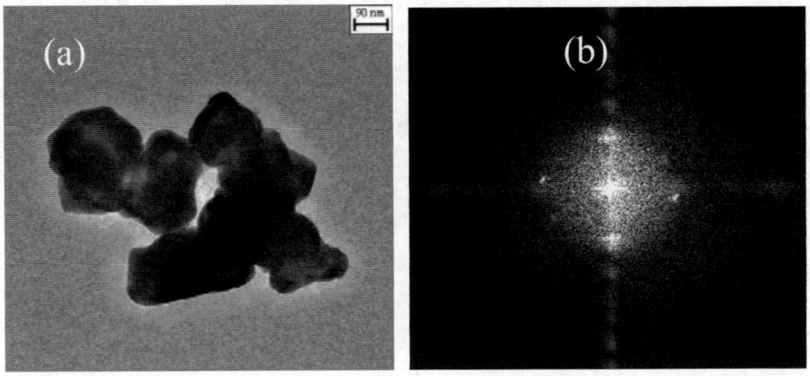

Fig. 4. TEM micrographs of (a) and SAED pattern of $Nb_2Zr_6O_{17-x}N_x$ as synthesized catalyst

The structural morphology of as-synthesized $Nb_2Zr_6O_{17-x}N_x$ catalyst has been carried out with TEM and selected-area electron diffraction (SAED). Figure 4a shows a low-magnification TEM image of $Nb_2Zr_6O_{17-x}N_x$ sample showed prismatic pseudo orthorhombic shaped particles clear edges with an average particle size of 100 ± 15 nm which was also observed in SEM results. This indicates the only a few nitrogen atoms was replaced with some of the oxygen of the mother materials oxide hence the geometry and shape are retained in the oxynitride samples. The SAED pattern taken from the same crystalline particle is shown in Fig. 4b, which showed the agglomerated crystalline behavior and also the calculated d-values from the pattern are well matched with the XRD results.

3.5 Elemental analysis By XPS

Fig. 5 shows the survey by XPS spectra of as-prepared oxynitride ($Nb_2Zr_6O_{17-x}N_x$) sample powder. XPS measurements showed the incorporation of nitrogen into the sample as elemental signal. The binding energies obtained in the XPS analysis are corrected according to the standard value of C ls. No peaks of other elements except C, O, N, Nb, and Zr are observed on the wide-scan XPS picture, indicating the high purity of the product. The peak area of these high-resolution scans was measured and is used for calculation of ratio between O and N in the product. The quantification of the peaks shows that atomic ratio of the O: N is19.66:1 indicates the doping of nitrogen in the sample. Since the compound has showed the absorption in the visible region (Fig. 2b) hence further the photocatalytic study of sample powder was carried out to decomposition of H_2S.

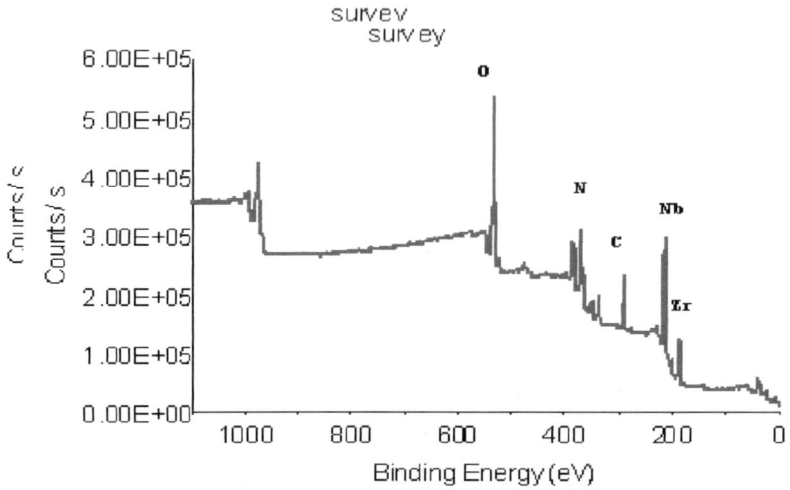

Fig. 5. Elemental survey of $Nb_2Zr_6O_{17-x}N_x$ by XPS-analysis.

3.6 Photocatalytic activities of $Nb_2Zr_6O_{17-x}N_x$ *catalysts*

Photocatalytic activity of $Nb_2Zr_6O_{17-x}N_x$ was investigated for the decomposition of hydrogen sulphide to generate hydrogen under visible light irradiation. Oxynitride samples were doped with Pt and RuO_2 and their photocatalytic activity was also studied under the identical reaction conditions and the data is summarized in Table 1. The undoped catalyst has shown excellent photocatalytic activity than the doped catalyst. However, the presence of Pt and RuO_2 had little or opposite effect, the one of the reason for decreased activity may be due to nitrogen leaching during the thermal treatment of doping process. The rate of evolution hydrogen to undoped and doped with Ni and RuO_2 photocatalyst was found to be stable. This behavior can be attributed to the non-deposition of sulfur on the surface of the photocatalyst due to the alkaline medium. It is known in the literature that in an alkaline medium sulfide forms soluble disulfide ions; hence no sulfur precipitation occurred during the photocatalytic process. The photochemical reaction for the photodecomposition of H_2S is followed as per the mechanism reported in our earlier communication (16). The undoped catalyst powders gave 13.5% quantum yield (the number of photons utilized for a desired chemical reaction divided by the number of photons absorbed by the catalyst) for the H_2 evolution at 550 nm, which is highest among all the photocatalyst reported in the present study. From the result it showed that

$Nb_2Zr_6O_{17-x}N_x$ is an efficient self-deposited photocatalyst to the decomposition of hydrogen sulphide to hydrogen.

Table 1: Photocatalytic evolution of H_2 by the decomposition of H_2S^a

Catalyst name	Band gap (eV)	H_2 (μmole/h)	Quantum yield (%)
1. $Nb_2Zr_6O_{17-x}N_x$	2.14	4283	13.5
2. $Nb_2Zr_6O_{17-x}N_x^b$	2.14	3446	10.8
3. $Nb_2Zr_6O_{17-x}N_x^c$	2.14	2766	8.7

[a]Catalyst, 0.5g; 250mL (0.5M aqu.KOH); H_2S (2.5mL/min); 450 W Xe lamp (Oriel) with cutoff filter ($\lambda \geq$ 420nm), [b] Pt and [c]RuO_2 doping

Conclusion:

Thermal ammonolysis of $Nb_2Zr_6O_{17}N$ at 1073 K gave a new Niobium Zirconium oxynitride visible light photocatalysts. The undoped photocatalyst showed high activity in the decomposition of hydrogen sulfide to the evolution of hydrogen as compared to the Pt and RuO_2 doped. These are our preliminary results and further modifications in the structure of the photoactive materials are in progress to improve the photocatalytic activity for the decomposition of water and H_2S for hydrogen production.

Acknowledgements

This research (paper) was performed for the Hydrogen Energy R&D Center, one of the 21st Century Frontier R&D Program, funded by the Ministry of Science and Technology of Korea.

References

1. Andew Mills, and Stephen Le Hunte, A overview of semiconductor particles J. Photochem. Photobiol. A: Chemistry ,Vol. 108, 1997, p. 1.

2. H. Kato, and A. Kudo, "Visible-light responseand photocatalytic activity of TiO_2 and $SrTiO_3$ photocatalyst codoped with antimony and chromium". J. Phys Chem B. Vol. 106, 2002, p.5029.

3. T. Ishii, H. Kudo, and A. Kudo, "H_2 evolution from aqueous methanol solution on SrTiO3photocatalyst codoped with tantalum and chromium ions under visible light irradiation". J. Photochem. Photobiol. A. Vol. 103, 2004, p.181.

4. T. Ishii, H. Kato, H. Kobayashi, and A. Kudo, "H_2 evolution from aqueous methanol solution on $SrTiO_3$photocatalyst codoped with tantalum and chromium ions under visible light irradiation". J. Photochem. Photobiol. A. Vol. 163, 2004, p.181.

5. S. A. Naman, "Photoproductionn ofhydrogen sulfide in vanadium sulfide collidal suspension : effect of temp and pressure". Int. J. Hydrogen Energy, Vol. 22, 1997, p.783.

6. D. Park, and J. O. Beag; "ZnS Photocatalyst, preparation thereof and merthod for producing hydrogen by use of the same", U.S. Patent No. 6,297,190 B1 (Oct. 2, 2001).

7. D. Park, and J. O. Beag; "CdZnMs photocatalyst including cations for water decomposition and preparation thereof and method of producing hydrogen by use of the same". U.S. Patent No. 6,517,806 B1 (Feb.11, 2003).

8. H. Chang, K. Kong, Y. Choi, E. In, Y. Choi, J. O. Baeg, and S. Moon, "Electronic structure of InTaO4 : A promising photocatalyst" Chem. Phys. Lett. Vol. 398, 2004, p.449.

9. A. Kudo, K. Omori, and H. Kato, "A novel aqueous process for preparation of controlled crystal formation and highly crystalline $BiVO_4$ powder from layered vanadates at room temperature and its photocatalytic and photophysical properties" J. Am. Chem. Soc. Vol. 121, 1999, p.11459.

10. H. Tokunaga, H. Kato, and H. Kudo, "Selective preparation of monoclinic and tetragonal $BiVO_4$ with sheet like structure and their photocatalytic properties", Chem. Mater. Vol. 13, 2001, p.4024.

11. H. Kato, H. Kobayashi, and A. Kudo, "Role of Ag+ in the band structure and photocatalytic properties of $AgMO_3$ (M : Ta , Nb) with pyrovskite structure" J. Phys. Chem. Vol. 106, 2002, p.12441.

12. A. Ishikawa, T. Takata, J. Kondo, M. Hara, K. Kobayashi, and K. Domen, "Oxysulfide Sm_2 $Ti_2S_2O_5$ as a stable photocatalyst for water oxidation and reduction under visible light (650nm)", J. Am. Chem. Soc. Vol. 124, 2002, p.13547.

13. A. Kasahara, N. Nukumizu, T. Takata, J. Kondo, M. Hara, H. Kobayash, and K. Domen, "$LaTiO_2N$ as a visible light(600nm) driven photocatalyst(2)", J. Phys. Chem. B, Vol. 107, 2003, p.791.

14. M. Hara, T. Takata, J. N. Kondo, K. Domen, " Photocatalytic reduction of water by TaON under visible light irradiation", Catalysis Today, Vol. 90, 2004, p.313.

15. I. Tsuji, H. Kato and A. Kudo, "Visible-light-induced H_2 evolution from an aqueous solution containing sulfide and sulfite over a ZnS-$CuInS_2$-$AgInS_2$ solid solution photocatalyst" Angew. Chem. Int. Ed., Vol. 44, 2005, p.3565.

16. B. B. Kale, Jin-OoK Baeg, Jin S. Yoo, Sang Mi Lee, Chul Wee Lee, Sang-Jin Moon, Hyunju Chang, "Synthesis of a novel photocatalyst, $ZnBiVO_4$, for the photodecomposition of H_2S" Can. J. Chem., Vol. 83, 2005, p.527.

Down-hole Combustion Method for Gas Production from Methane Hydrate

Marco J. Castaldi *, Yue Zhou and Tuncel M. Yegulalp
Dept. Earth & Environmental. Eng.
Henry Krumb School of Mines
Columbia University
New York, NY USA
mc2352@columbia.edu

Abstract

As global energy demands increase, there will be a critical need to employ fuels that have minimal environmental impacts. While hydrogen is considered the ultimate fuel for that purpose, hydrocarbon based fuels will be in use for a considerable time and will probably be the first major source of hydrogen for next generation power systems. The best fuel for this transition will be natural gas because of the high hydrogen to carbon ratio. Natural gas hydrates have the potential to meet the global energy needs for the foreseeable future. The amount of energy stored in natural gas hydrates is significantly more than that of all the other fossil fuels combined. However, most proposed techniques for recovering the natural gas from the hydrate require too much energy input, raise safety concerns or can only recover a fraction of the gas. This paper discusses the technical feasibility of a novel method of gas production in hydrate reservoirs. The method is an in-situ process that creates a point heat source in the hydrate formation. This method not only tremendously reduces the energy input, but also has the potential to sequester as much or more CO2 than would be produced from burning the natural gas produced, thus making this in-situ methane production technique carbon neutral. Calculations and simulations using FEMLAB™ show the method offers an energy efficient and environmental friendly way to produce natural gas from hydrate reservoirs. Preliminary results indicate that it is possible to have an energy efficiency of near 90% for land based reservoirs for hydrate loadings of 20%. The simulation results also suggest that alternatives to a point source can be employed which may improve the process efficiency.

Keywords: methane hydrate, hydrate dissociation, CO2 sequestration, alternative energy

INTRODUCTION

Interest in methane hydrates as a potential source for clean hydrocarbon energy supply for the world has been steadily increasing. As a result, numerous researchers have explored different aspects of hydrates, from fundamental properties to extraction and production potential. An extensive collection of fundamental and practical developments related to gas hydrates can be found in Sloan (Sloan, 1998). Most of the proposed extraction scenarios require too much energy input (low energy efficiency), raise safety concerns or can convert only a part of hydrate to gas in a reservoir (small levels of resource recovery), or suffer from some combination of these deficiencies, and do not consider CO_2 sequestration as part of the production process. CO_2 sequestration as hydrates has also been to subject to discussion in the literature, mainly as part of CO_2 sequestration in deep oceans, usually independent of methane hydrate extraction.

In this paper we introduce a novel concept of generating a point heat source, via in-situ combustion, strategically located in the hydrate deposit which enables methane production using only about 10% of the energy content of the reservoir for the process. This combustion process is capable of utilizing CO_2 as a diluent to adjust adiabatic temperatures to the point where methane hydrates decompose, yet CO_2 hydrates still can form and replace methane. Feasibility calculations were made for using this point heat source as a temperature-balancing extraction process by which methane gas can be efficiently produced, while sequestering CO_2 in a stable environment. Our continuing research effort seeks to gain an in-depth understanding of the processes involved in using in-situ combustion to locally heat a portion of the hydrate reservoir in order to dissociate methane hydrate and produce methane gas with minimum loss of energy present in the source.

BACKGROUND

The U. S. presently imports nearly half of its petroleum needs. With domestic reserves being a small fraction of worldwide reserves, domestic production cannot be significantly expanded. Natural gas, which could substitute for petroleum in many uses, is also imported. "At present, the United States relies on the accelerated use of natural gas to simultaneously achieve aggressive economic and environmental goals. Fundamental to this strategy is the presence of abundant and affordable supply of domestic natural gas. However, there are increasing concerns about the long-term certainty of this supply" (NETL, 2004)

Domestic as well as global reserves of natural gas in the form of methane hydrates are projected to be enormous (NETL, 2004, Dillon, 1992). In fact, methane hydrate deposits throughout the world have been estimated to represent by far the greater portion of the world's fossil energy reserves, greater than the world total coal and petroleum reserves combined. The U.S. Department of Energy estimates that for the U.S. alone, hydrates represent some 200,000 Trillion cubic feet (Tcf) of the total 227,500 Tcf of known natural gas resources. In fact the hydrate deposits, occurring at great depths primarily in the oceans as well as in arctic regions, dwarf the total known combined oil and non-hydrate gas reserves.

Unfortunately, although it has been demonstrated that methane can be recovered from dissociated hydrates, no economic method yet exists. Therefore, these huge reserve estimates can only be classified as "inferred" sub economic reserves or just resources.

Three potential production methods are recognized in the literature (Holder, et al., 1984). These are (1) thermal stimulation, (2) pressure reduction, and (3) inhibitor injection. Of these the only one which is likely to lead to large-scale economic production is thermal stimulation. It is thermodynamically feasible to supply the necessary energy if the freed methane is burned, because the heat of dissociation is only about 10% of the heat released by methane combustion excluding heat losses. In-situ combustion would minimize heat losses but such combustion may not be stable in high water content strata. Another possible thermal stimulation method could be to pump a heated fluid, such as steam or hot CO_2, to the desired location. However, pumping a heated fluid from the surface results in such a high heat loss that essentially all of the heating value of the recovered methane is used to supply the needed energy. In fact, for deep deposits all the heat can be lost in transit. The primary production challenge is to devise a technology for delivering heat throughout hydrate reservoirs without losing heat to non-productive areas or overburden.

The pressure reduction method involves reducing the reservoir pressure to a value below the hydrate dissociation threshold at the prevailing temperature. However, the dissociation energy must still be supplied and thus the formation temperature drops requiring ever lower pressures for dissociation. Thus, this approach typically requires mining the solid hydrates and pumping a slurry to the surface. Such a slurry mining system has yet to be demonstrated to be economically and technically feasible.

The use of inhibitors has gained interest lately because of the favorable economics. In addition, they are used in the field as hydrate control methods during production (Komai, et al., 2000). This method however, requires the injection of a chemical into the hydrate zone and must be evaluated in light of a number of issues. For example, not only performance and environmental impacts need to be assessed but reclamation of the chemical, space constraints (especially for offshore platforms), and maintenance must be factored in as well.

It has also been suggested to pump carbon dioxide to displace methane from the hydrates by formation of carbon dioxide hydrates. However this reaction is limited because the reaction is slow at temperatures at which methane hydrates are stable. Even though the formation of carbon dioxide hydrates would release some energy in displacing the methane from the hydrate, the energy would not be enough to overcome the pumping energies or production of methane at any significant rate. As shown in literature (Smith et al., 2001), the heat of formation for CO_2 hydrate is higher than for methane hydrate in porous media. However, if some external energy was provided to first begin the methane hydrate dissociation, followed by initiation of the CO_2 hydrate formation, then it would be feasible to accomplish replacement of methane by CO_2.

The fact that CO_2 hydrate forms at slightly higher temperatures than where CH_4 hydrate decomposes enables one to implement the concept of temperature balancing by using a point heat source. That is, only if the heat source is applied locally and can be maintained to provide the right temperature rise within the formation. The use of liquid oxygen and liquid fuel will widen the flammability regime enough to maintain sufficient reaction rates at the desired moderate temperatures. The combustion can be further moderated by co-injection of CO_2 which will also help stimulate CH_4 hydrate decomposition.

Theoretically this idea will work because at a given pressure, carbon dioxide hydrate formation can occur at a slightly higher temperature than that for methane dissociation,

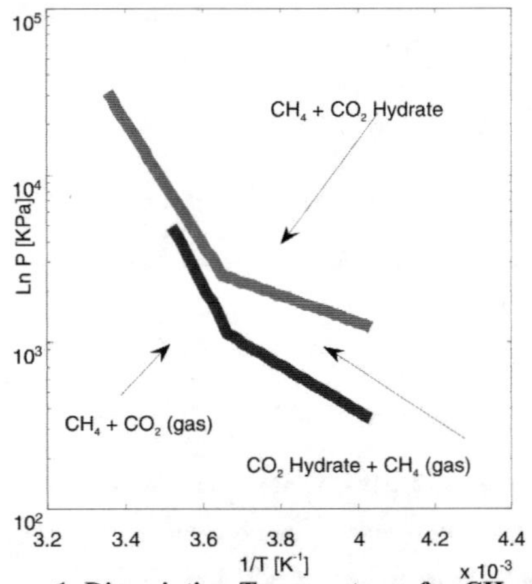

Figure 1. Dissociation Temperatures for CH_4 and CO_2. Based on data and tables from Kamath (1984) and Sloan (1988).

(Kamath, 1984 and Sloan, 1998) (Figure 1). Note that the slope of the CO_2 line is steeper than the methane line. Thus, the kinetic rate of carbon dioxide formation is higher at a temperature near the methane dissociation temperature than below it. Reformation of methane hydrate is thus minimized. Further, a mixture of oxygen and carbon dioxide may be used as the oxidant rather than air, with added carbon dioxide (rather than water supplied to control the injected fluid temperature). Not only can nitrogen dilution of the produced gas be eliminated but carbon dioxide is also sequestered. Thus the inclusion of CO_2 to the flow not only enables the potential for sequestration of the CO_2 into hydrate but also reduces required methane fuel usage and water injection, further boosting energy efficiency. The amount of CO_2 sequestered by this approach will determine the degree at which the product (CH_4) may be considered carbon-free.

PRELIMINARY CALCULATIONS

In order to establish the limits of energy efficiency that can be expected from the use of a point heat source for dissociating methane hydrates we performed a series of calculations. Figure 2 shows the results of a preliminary calculation done to determine the amount of energy input needed to produce a certain amount of energy contained in methane gas. The efficiency (η) shown in Figure 2 is defined as [energy recovered – energy input]/energy recovered. This energy balance clearly shows that only a small amount of energy is needed if the heat required decomposing the hydrate into methane gas and liquid water is applied locally. More importantly, the reservoir does not have to be completely saturated with methane hydrate. To the contrary, the calculation shows that even if only 1% of the formation contained methane hydrate, and the temperature needed to be raised by 10°C, one would need 70% of the energy that would be contained in the methane produced. While that is an extreme case, it shows the potential of this technique. It is more likely that the amount of hydrate in the formation would be closer to 10-15% (Kastner, et al., 1995) and

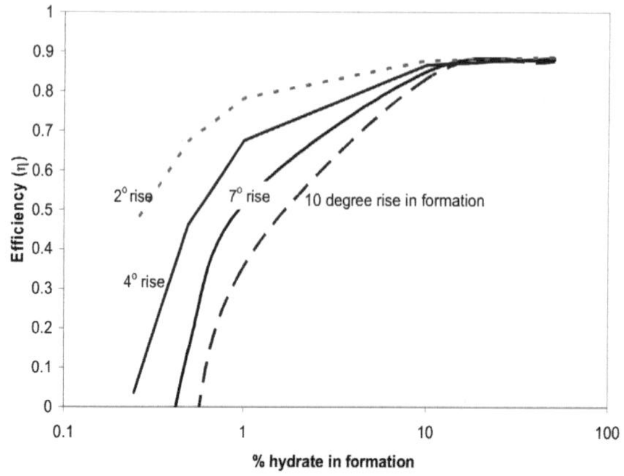

Figure 2 Calculation showing that heat applied locally will require a fraction of the energy that can be produced from the hydrate bearing formation.

therefore the amount of energy required drops significantly. For example, if the amount of hydrate in the formation is 10% and the temperature needs to be raised by 10°C, only 18% of the liberated energy would be required or the efficiency would be 82%. If the heat source is positioned properly, it is likely that a smaller temperature rise, about 5°C, would be needed, thus raising the efficiency.

MODEL SETUP

A two dimensional model was set up to simulate the heat and mass transport from the combustion point to the surrounding system. The model tracked the temperature front of the heat movement through the entire formation and was coupled to a decomposition rate equation taken from literature. The main objective was to track the temperature front at which the hydrate decomposes to liquid water and pressurized methane gas. However, it is recognized that as the temperature rises to a point where hydrate decomposition begins, there will be a condition imposed in the model such that a steady temperature is maintained until all of the phase change is complete. The rate equation used is expressed on a per unit hydrate mass basis which provides versatility in terms of exploring different amounts of hydrate in the formation. To simulate the time lag associated with complete hydrate decomposition, a Heavysides function has been incorporated into the model. This function provides the ability to adjust the decomposition time based on the amount of hydrate present. For example, if a very low hydrate loading is present, it is conceivable to imagine the time required for hydrate decomposition is very fast compared to the movement of the temperature front, thus the Heavysides function would emulate a step function. Alternatively, if the hydrate loading were high, the Heavysides function would emulate more of an S-shaped curve with respect to time and may then be on the order of the temperature boundary movement time.

A schematic of the hydrate reservoir system used is shown in Figure 3. The system comprises of three main zones. The first zone was considered permafrost or overburden. The second zone was the hydrate layer where the heat source would be placed. The third zone considered was the free gas zone. This is the zone, where many propose to drill to and produce gas by simply relieving the pressure. This process would work, but the length of production time, and hence the ultimate amount of gas recovered is uncertain. The primary limitation is the fact that as the gas expands from the pressurized free gas zone to the well the temperature will drop, thus establishing conditions to form hydrates which will limit or stop the gas flow.

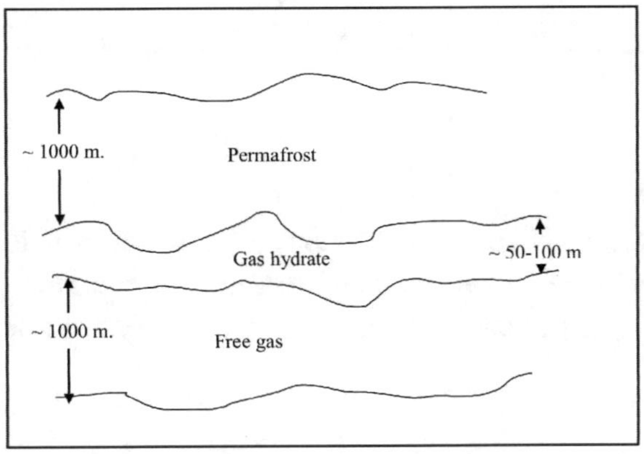

Figure 3. Conceptual schematic of matrix used in FEMLAB® simulation.

We are investigating whether to place the heat source at the hydrate and free-gas boundary or somewhere within the hydrate zone. A first order analysis indicates the heat source should be placed at the hydrate and free-gas boundary. This is the position that will require the least amount of energy to maintain a steady gas flow through the well. The concept relies on the principle that once the depressurization occurs near the well, only a small amount of energy is required to maintain the temperature surrounding the well a few degrees above the decomposition temperature.

The model was initially set up with the heat source in the center of the hydrate region to take advantage of symmetry and avoid a moving boundary problem. This enabled us to determine the feasibility of producing methane from local heat source concept. A next generation of the model currently being developed will put the heat source in an optimal position within the matrix and address the moving boundary issue. The model has been developed with few assumptions and employed the fundamental laws of mass and energy conservation and Darcy's law.

Consider a uniform distribution of hydrate in the porous media with initial temperature T_0 and pressure P_0. The heat source is located in the center of the hydrate layer. The heat source was kept as constant temperature T_c, which is the combustion temperature based on an adiabatic calculation of stoichiometric amounts of diesel fuel, liquid oxygen and liquid CO_2. The temperature can be adjusted by changing the equivalence ratio (methane to oxygen ratio) or the amount of CO_2 or both. A gas collection well is located under the heat source in the free gas zone and has been set at a pressure of around 2 bar based on a calculation using aerodynamic head considerations.

Each zone was characterized using literature values specific to each. For example, the porosity of the permafrost was considered to be about 1000 times less then the porosity of the free gas zone. The porosity of the hydrate zone was initially set to a value near that of the permafrost, then was reset once hydrate began to decompose to a value close to the free gas region. Most values were taken from Sloan (Sloan, 1998).

The governing equations employed in the simulation are provided below. The heat transfer equation is given by a time dependent diffusion, convection and conduction heat equation.

$$\rho C_p \frac{\partial T}{\partial t} + \rho_g C_{pg} v \nabla T - \nabla \cdot (K \nabla T) = Q \tag{1}$$

Where k is the thermal conductivity, ρ is the density of the hydrate formation; Cp is the specific heat capacity of the hydrate formation, K is the heat conductivity of the hydrate formation, ρg is the methane gas density, Cpg is the specific heat capacity of methane gas and υ is the methane gas velocity.

The methane gas velocity v in Equation (1) is calculated from Darcy's Law shown below. The velocity is determined by the pressure gradient that is established from the decomposition of hydrate which is occurs from a temperature increase. In additional, the fluid viscosity and the structure of the porous media all change as the heat front moves through the media.

$$v = -\frac{k}{\mu}\frac{\partial P}{\partial x} \tag{2}$$

Here k is the permeability of the porous media, which is different in the hydrate zone compared to the decomposition zone. Currently this change is handled using a Heavside Function to switch between k values. It is anticipated that a more descriptive, physically significant correlation will be developed during the experimentation. Lastly μ is the methane gas viscosity which is a function of temperature and impacts the velocity of the mass front. Therefore, we have a coupled set of equations in (1) and (2).

Darcy's law describes the flow in porous media with the pressure gradient as the only driving force. Actually, in the real situation, the gas close to the collection wells experiences a much larger velocity gradient than the gas dispersed throughout the media. As a result, the transport of mass by shear stress in the fluid is also of importance. Therefore a Brinkman equation, shown below, should be used to calculate the gas velocity in the region, but this is intimately tied to the velocity and the velocity gradient which is impacted by equation (1) in a highly non-linear fashion.

$$\nabla P = -\frac{\mu}{k} v + \mu \nabla^2 v \qquad (3)$$

In addition, the methane gas velocity is also needed for the mass balance calculation below.

$$\nabla(\rho_g v) = \frac{-\partial(\phi S \rho_g)}{\partial t} \qquad (4)$$

In this equation ϕ is the porosity of the matrix which changes as the methane hydrate decomposes, S is the saturation of the methane hydrate, υ is the velocity and ρg is the density.

Referring back to the heat equation (1), the Q term represents the latent heat absorbed while the hydrate changes phase. As the hydrate decomposes, the Q term represents a heat sink which is the energy required for phase changing only, not raising the temperature of the matrix. After the phase change occurs, there is no heat sink so the Q term is zero. In addition, this term is directly related to the loading of the methane hydrate. The larger methane hydrate loading, the bigger the value of Q.

Overlaid on the dynamic, coupled heat and mass transfer equations of the system is the equilibrium relationship of hydrate decomposition. Equation (4) shows the thermodynamic equilibrium relationship between the hydrate decomposition temperature and pressure, which is used in other modeling studies.

$$P_D = \exp(A - B / T_D) \qquad (5)$$

PD is the gas pressure at the dissociation interface and TD is the methane hydrate dissociation temperature. A and B are empirically determined constants. The impact of this relationship on the model fit of the experimental data will be investigated. An anticipated outcome of this SGER will be to gain some understanding of this.

From the discussion, the unknown coupled functions are: T(x, t), P(x, t), ρg (x, t), and υ(x, t). The pressure and temperature changes will affect the methane hydrate decomposition rate, which result in a different heat transfer rate.

RESULTS

Presented below are the simulation results from the temperature, pressure and velocity profiles for the steady state conditions. Also depicted in the Figures is a simulated well bore that would be used to extract the methane gas. As described above, the pressure of the well bore was calculated to be near 2 bars, thus creating a pressure sink and establishing the desired gradient for gas flow from the decomposed hydrate and the free gas. The surrounding pressure around the well bore is lower than the far field pressure. This condition allows for the gas released in the hydrate to migrate toward the well, thus allowing the entire hydrate zone that is decomposed to drain. Since the flow is highly restricted in the upper sediment layer and the regions of stable hydrate the desired flow pattern is established. The maximum flow velocity is through the decomposed hydrate and free gas region 0.008 m/s, which is two-orders of magnitude higher than the upper limit of fluid velocity in the overburden and stable hydrate zones. This higher velocity is attributed to the high-pressure gradient at the heat source. Figure 4 shows the temperature profile for the entire calculation domain to provide a sense of the size of the local heat source in relation to the affected region. The scale of the simulation should be noted at this point. The horizontal scale was taken to be 500 units and the vertical scale was taken to be 2000 units with the hydrate layer being about 100 units. The heat source is only 1 unit in diameter.

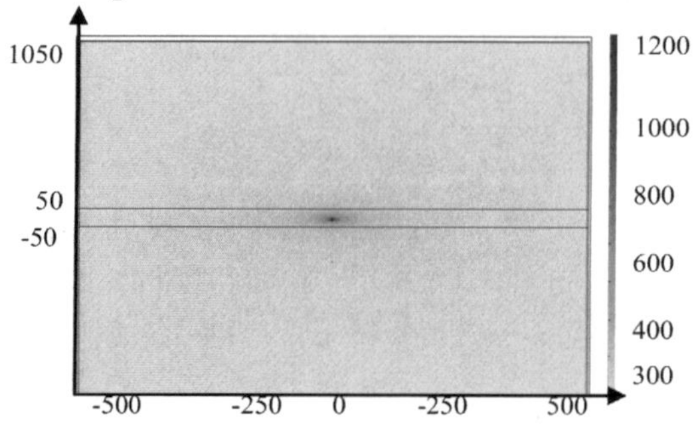

Figure 4. Temperature Distribution

These units can be translated into meters if desired and gives an idea of the influence such a small heat source has in the overall matrix

The pressure distribution and flow pattern are shown in the Figures 5 and 6. The point heat source was simulated as a constant energy input as described above. Very near the heat source, the pressure is high due to the conditions that are setup by this process. Not only has the hydrate decomposed completely, but the methane gas and water are now being superheated and thus result in a local pressure buildup. This pressure will not reduce until gases are withdrawn from the well bore, positioned in the free gas zone in this simulation. Because the permeability is small compared to the free gas zone that is below the hydrate layer, it is hard for the released the gas to get through to the overburden and as a consequence, the pressure builds up at the interface between hydrate layer and the upper sediments. This can bee seen in Figure 5.

In Figure 6, arrows represent the velocity profile of the methane gas only. The length of the arrows are proportional to the velocity. In the center of the heat source the gas maintains a fairly high temperature and thus sets up a steep pressure gradient, so the gas velocity is high. Moving away from the heat source, the temperature drops which results in

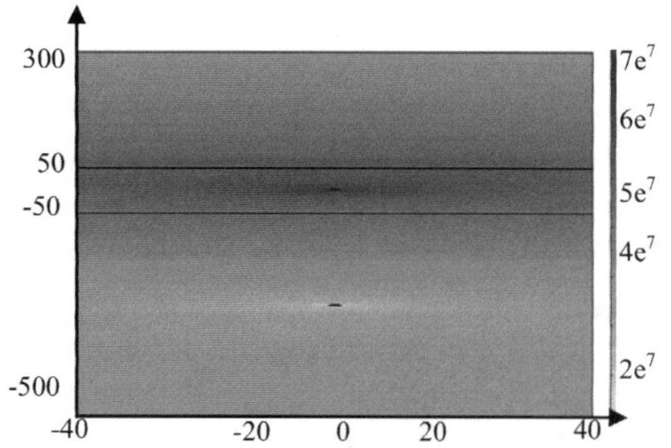

Figure 5. Pressure Distribution

a commensurate drop in pressure. As the temperature drops below the hydrate decomposition temperature, the pressure gradient becomes negligible and gas flow essentially stops. This is considered the far field condition, which is maintained throughout the simulation.

The velocity distribution of the methane gas is shown in Figure 6. The flows that are established result from the pressure gradients that are set up from the local heating and dissociation of the hydrate. The right hand scale is velocity in meters per second. The interesting thing to note is the region of influence the small point heat source has throughout the matrix. Although the point heat source is only 1 unit in diameter, it is causing methane gas movement 40 units in either direction.

Now one can see why applying a local heat source at the properly located position within the matrix is so efficient. Not only is all the energy required to dissociate the hydrate release where it is needed, but also the far reaching impacts the source has make it very

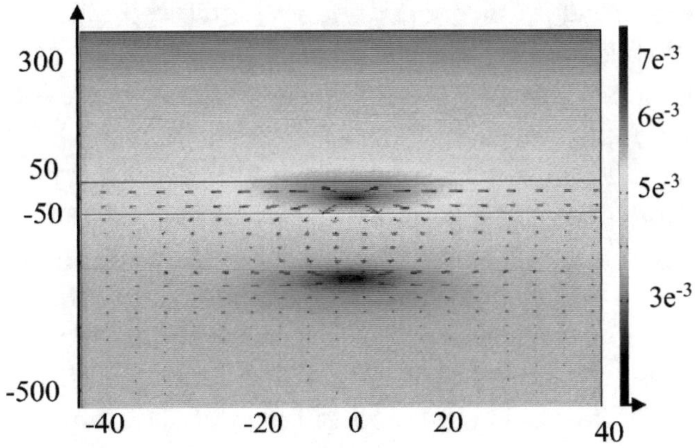

Figure 6. Velocity Distribution

effective. Not shown in Figure 6 is the far field region where the velocity goes to zero because the temperature is at the reservoir temperature and the pressure gradients are extremely small.

CONCLUSIONS

In this paper we have shown results of feasibility calculations showing the potential to produce methane gas from hydrates in ground formations using a local heat source. It is possible, using this process to produce methane with an energy efficiency of nearly 80%. Heat transfer simulations have been completed that provide information on the types of temperature gradients that can be expected during this type of operation.

The simulation results give us a preliminary understanding of the down-hole combustion process, which has the potential of being implemented efficiently, and safely to produce methane from methane hydrate formations. We have observed that:

1. The high temperature created by local combustion establishes a reasonable temperature gradient in the hydrate layer within a feasible time.

2. The heat flux supplied by high temperature allows continuous dissociation of hydrate.
3. The high temperature of the combustion also enhances the gas flow in the porous media, which will in turn helps to improve the heat transfer and gas collection.

EXPERIMENTAL WORK

To assess the potential for a viable gas production process based on in-situ controlled combustion coupled with CO_2 sequestration, we developed a three- phase experimental procedure:

Construction of the test apparatus

We completed the design and building of a test apparatus. This apparatus consists of a chamber where hydrate is formed by adjusting temperature and pressure. To simulate hydrate bearing formations commonly seen in various exploration sites including Beaufort-Mackenzie Basin, fine grained (~1mm dia.) silica and water mixture is to be used. The temperature and pressure are continuously monitored and regulated in order to replicate in-situ thermodynamic equilibrium conditions. Initially a point heat source will be imbedded in the middle of the hydrate block. During the first phase, the local heat source will be electrical. In the subsequent phases of the experiment, this will be replaced by a pipe through which oxygen and CO2 mixture will be injected to provide targeted stoichiometric mixtures for combustion for local heat generation. A series of various mixures will be explored to find optimal conditions for combustion, temperature, production rates and CO2 sequestration. Near the heat source there will be a number of production pipes at varying distances. As heat is applied to the middle of the hydrate formation, the local dissociation will take place. Gas flow will be monitored in the production pipes at various controlled outlet pressures. Water with gas will also is expected to flow out. This water will be returned to the system in order to maintain the mass balance and simulate what may have to be done in the field. In the experiments make-up water will be injected from the water supply line.

This apparatus will be used for our first phase test. As shown in Figures7, the device consists of a chamber where hydrate is formed by regulated temperature and pressure. The chamber will be filled with the sand-pack porosity from 30 to 45%, which is a typical range for the permafrost region in Alaska (Bradley, 1989). The methane gas and water is fed into the chamber from the bottom of the chamber.

Figure 7. Test apparatus external view

On the top of the chamber we inserted a number of tubes that will house a thermocouple and work as gas sample collectors. In order to measure the distribution of gas flow and the temperatures, the thermocouples and gas sample probes are uniformly distributed on the radial and circumferential directions. As shown in the sectional view (Figure 8) of the chamber below, probes are also located at different depths along the axial direction.

Using this experimental apparatus it will be possible to

- Verify the feasibility of gas hydrates dissociation by down-hole fire flooding.
- Understand important characteristics such as temperature and pressure profiles for methane hydrate formation and dissociation in pure water system as a reference.
- Obtain the relevant data to develop the models for methane hydrate dissociation by combustion method.
- Maintain constant pressure for the duration of the experiment to approximate far field conditions.

References:

NETL, (2004) http://www.netl.doe.gov/scng/hydrate/rd-program/rd-program.htm

Sloan, E. D., (1998), Clathrate Hydrates of Natural Gases, 2nd Ed., Marcel Dekker, Inc. New York, 1998

Dillon, William (1992), U.S. Geological Survey, Gas (Methane) Hydrates -- A New Frontier

Holder, G. D., Kamath, V. A., and Godbole, S. P.,(1984), The potential of natural gas hydrates as an energy source: Annual Review of Energy, v. 9, p. 427-445.

Kamath, V. A., (1984), Study of Heat Transfer Charactristics During Dissociation of gas Hydrates in Porous Media, Ph.D. Dissertation, U. of Pittsburgh, University Microfilms No. 8417404, Ann Arbor MI.

Kastner, M., Kvenvolden, K.A., Whiticar, M.J., Camerlenghi, A., Lorenson, T.D., (1995) Proc. Ocean Drill Program, Leg 146, 175.

Komai, T, Yamamoto, Y and Ohga, K, (2000) "Dynamics of reformation and replacement of CO_2 and CH_4 gas hydrates", Gas Hydrates: Challenges for the future, Annals of the New York Academy of Sciences, vol 912.

Smith, D., Seshadri, K., Wilder, J.W., (2001) "Assessing the Thermodynamic Feasibility of the Conversion of Methane Hydrate into Carbon Dioxide Hydrate in Porous Media", Proceeding of the First National Conference on Carbon Sequestration, National Energy Technology Laboratory (NETL) Conference, Poster Presentation

Evaluation of Process Variable Interaction in the Biodiesel Production from Castor Oil Ethanolise

Silva Lima N.*, Wolf Maciel M.R., Batistella C.B., Maciel Filho, R.

Laboratory of Separation Process Development (LDPS), School of Chemical Engineering, State University of Campinas, São Paulo, Brazil, CP 6066, CEP 13081-970. Phone: 55 19 37883971 email*: niveals@feq.unicamp.br

Abstract

Biodiesel is an alternative biodegradable and nontoxic fuel, which is essentially free of sulfur and aromatics. It is usually produced by a transesterification and esterification reaction of vegetable or waste oil respectively with a low molecular weight alcohol, such as ethanol and methanol.

This work presents the transesterification process of refined and pharmaceutical castor oil with ethanol in the presence of sodium ethoxide and sodium hydroxide as catalysts, because they present better conversion and smaller reaction time. Chemically, this oil contains approximately 90% of ricinoleic acid, what gives to the oil some important characteristics, such as its ethanol solubility at 30°C, allowing reaction at low temperature.

The proposed process of biodiesel production was evaluated and optimal operating conditions range was identified by application of the factorial design and response surface methodology.

The studied variables were reaction temperature, catalyst concentration and ethanol : castor oil molar ratio. The experiments were carried out according to a 2^3 complete factorial design plus three central points and six axial points, called star points. High-performance size-exclusion chromatography, also called gel-permeation chromatography, was used to evaluate the influence of different variables affecting the transesterification. A second-order model was obtained to predict conversion as a function of temperature, catalyst concentration and ethanol castor oil molar ratio. The model has been found to describe the experimental range studied adequately. Good phase separation was obtained with the methodology developed. The best result for laboratory-scale reactions were obtained at 30°C and 1% of sodium ethoxide by weight of castor oil.

A small scale up was realized with a batch reaction, mechanical agitation and sodium hydroxide as catalyst. The effect of the scale up was interpreted by variations of ethyl ester conversions.

Keywords: Biodiesel, castor oil, ethanol, transesterification.

1. Introduction

Since the petroleum production is progressively decreasing, the demand for alternative energy is an important matter of study. Vegetable oil fuel, or biodiesel, is a potential substitute for diesel fuel because it is made from renewable sources.

Biodiesel is produced through transesterification reaction of vegetable oils and animal fats preferentially with alcohol of low molecular weight. Ethanol is an alternative to methanol, because it allows production of entirely renewable fuel. This reaction can be carried out in the presence of alkaline, acid or enzyme catalysts or, even using supercritical alcohol.

Castor oil has a proper characteristic that is its composition of 90% of ricinoleic acid (Figure 1). The castor oil is the best substance for producing biodiesel because it is the only one that is soluble in alcohol, and does not require heat and the consequent energy requirement of other vegetable oil in transforming them into fuel [1].

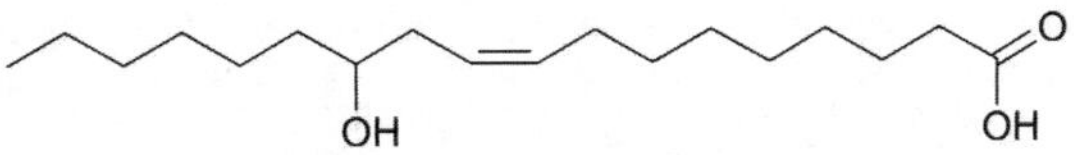

Figure 1 - Ricinoleic Acid

This work presents the transesterification process of refined and pharmaceutical castor oil with ethanol in the presence of sodium ethoxide and sodium hydroxide as catalysts, the reaction time was 90 minutes. In Lima Silva *et al.* (2006) [2] is presented the same process, however in that case the reaction time was 30 minutes, the sodium hydroxide was not used and the studies were performed in laboratory scale.

High-performance size-exclusion chromatography, also called gel-permeation chromatography, was used to evaluate the influence of different variables affecting the transesterification. This chromatographic technique permits the separation of the glyceride groups present, according to the molecular size [3].

Different variables affect the castor oil transesterification such as reaction temperature, ethanol : castor oil molar ratio , catalyst concentration , level of agitation and reaction time. Response-surface methodology (RSM) was used, because it allows the simultaneous consideration of many variables at different levels and the interactions between those variables, using a smaller number of observation than conventional procedures.

2. Experimental Procedures

2.1 Materials and Apparatus.

Commercial refined castor oil (Aboissa; Brazil) and pharmaceutical castor oil (Delaware; Brazil). The ethanol was anhydrous (Merck), the sodium ethoxide (Aldrich) was 95% pure and the sodium hydroxide (synth) was 97% pure.

The experiments were carried out in a 250 ml flask, connected with a condenser, the reaction mixture was agitated by magnetic stirrer and the After reaction, a rota-evaporator was used in order to recover the ethanol excess and the mixture was placed in a separation funnel. The water content was determined by Karl Fisher.

2.2 Reaction.

Initially, the commercial castor oil and pharmaceutical castor oil were used. Initially, 50 g of vegetable oil was added in a 250 ml flask and preheated to the desired temperature. After that, the catalyst was prepared by dissolving the sodium ethoxide in the desired amount of ethanol. This ethanolic solution was added to the castor oil and the reaction started at this time. The system was kept at normal pressure and the experiments were carried out at constant temperature. The agitation was kept constant. The reaction time was 90 minutes, and, during this time, samples were collected, diluted in tetrahydrofuran grade HPLC (Tedia), cooled instantaneously and analyzed in the HPSEC, evaluating the triglycerides (TG), diglycerides (DG), monoglycerides (MG), esters (EE) and glycerol (GL) contents, according to the methodology used by Filléres *et al.*, 1995 [4]. The conversion was obtained analyzing the corresponding sample at 90 minutes of reaction in the HPSEC. The conversion (Y) was defined by dividing the peak area of the ester by the sum of the peak area of all components.

$$\left(Y = \frac{A_{EE}}{A_{TG} + A_{DG} + A_{MG} + A_{EE}} 100 \right)$$

A typical chromatogram is presented in Figure 2. After the reaction, the mixture was evaporated in the rota-evaporator under vacuum up to constant weight and then transferred into a separatory funnel. Initially, smaller ethanol : castor oil molar ratios were used (6:1 – 9:1), however did not have phase separation, these reactions mixtures have got a kind of gel or homogeneous phase. Then all experiments were used large ethanol : castor oil excess which have indicated better phase separation. And after some time which depends on the composition of the mixture, and at room temperature two layers are formed: the upper one rich in ester and the lower one rich in glycerol, catalyst and some impurities. Later, the ester phase was separated.

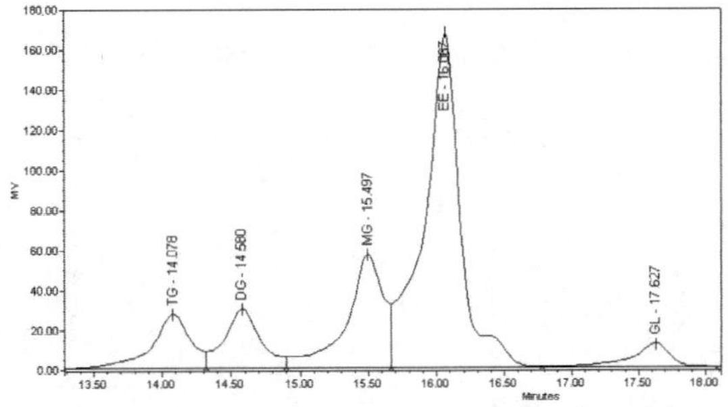

Figure 2 - Typical chromatogram of ethanolysis reaction using the HPSEC method:TG-Triglycerides; DG-Diglycerides; MG- Monoglycerides; EE- Ethyl ester; GL- Glycerol.

2.3 Batch reaction experiments

The pharmaceutical castor oil (500 g) was added in a batch reactor with mechanical stirrer and preheated to the desired temperature. After that, the catalyst was prepared by dissolving the sodium hydroxide in the desired amount of ethanol. In these experiments were realized a small scale up of the reaction. The ester conversion reached with pharmaceutical castor oil was similar of the commercial castor oil.

3. Results and Discussion

3.1 Experimental design

Response surface methodology (RSM), or response surface analysis was used because it allows the simultaneous consideration of many variables at different levels as well as the interactions between those variables, using smaller number of observations than conventional procedures. Furthermore, statistical-interference techniques can be used to assess the importance of individual factors, the appropriateness of their functional form and the sensitivity of the response of each factor [5,6]. Typically, the central point is replicated several times to provide an independent estimate of experimental error [7].

The response surface methodology was chosen to study the optimization of 3 selected factors: temperature (T), ethanol : castor oil molar ratio (E:O) and catalyst concentration (C). The experiments were carried out according to a 2^3 complete factorial design plus three central points and six axial points, called star points. The distance of the star points from the central point is given by $\alpha = (2^n)^{1/4}$, where n is the number of independent variables, for three factors $\alpha = 1.68$ [8]. Statistica software version 7.0 from Statsoft Co. was used in this study. Table 1 shows the decoding values for temperature, catalyst concentration and ethanol : castor oil molar ratio.

Table 1 - Decoding values of independent variables used in the experimental design

Level	Independent variable		
	Temperature (°C)	Catalyst (%)	Ratio: Ethanol/CO
$-\alpha$	30	0.5	12:1
−1	40	0.7	13.6:1
0	55	1	16:1
1	70	1.3	18.88:1
$+\alpha$	80	1.5	20:1

The factors (designated as X_i) were set independently of each other within the following limits: Temperature (X_1): 30°C ≤ T ≤80°C; catalyst concentration (X_2): 0.5% ≤ C ≤1.5%; ethanol : castor oil molar ratio (X_3): 12:1≤ E:O ≤ 20:1. Temperature level was chosen by considering the properties of the reactants. The upper temperature level, 80°C, was determined by the boiling point of ethanol; the lower level, 30°C, was the room temperature [9]. Catalyst concentration level was from 0.5% to 1.5%, according to literature data for other case studies [9, 10]. The range of ethanol : castor oil molar ratio was chosen based on our previous experiments, which have indicated better phase separation and reaction time.

The simplified RSM equations take into account only the significant coefficients. The coefficient b_o is the outcome (response) at the central point and the other coefficients measure the main effects and the interactions of the coded variable X_i on the response Y. The coefficient b_o is the outcome at central point and the other coefficients measure the mean effects and the interactions of the coded variables X_i on the response Y [11]:

$$Y = b_o + \sum b_i X_i + \sum b_{ii} X_i^2 + \sum b_{ij} X_i X_j \quad (with\ i \neq j) \qquad (1)$$

As shown in Figure 3, a plot of conversion to ethyl ester vs. time, the reaction is very fast, and conversions larger than 90% are reached within 15 min. However, the time was considered up to 90 min, since the reaction was stable. This result shows the optimization of the reaction time, what can lead to an alternative continuous process for biodiesel production, in contrast to the works presented in the open literature up to now.

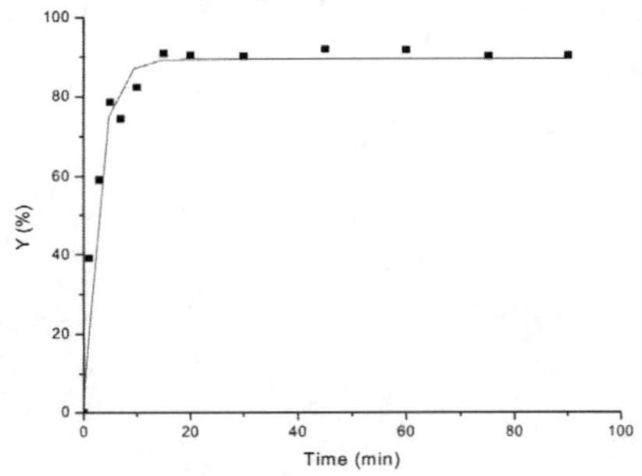

Figure 3 - Conversion to ethyl ester vs. time. Catalyst: NaOEt; T = 30°C; C =1 wt%; Ethanol:oil molar ratio - 19:1.

The effect of ethanol concentration on the ester conversion was studied by varying the ethanol : castor oil molar ration from 6:1 to 39:1 while maintaining the temperature and sodium ethoxide concentration at 30°C and 1 wt%, respectively. As shown in Figure 4, the ester conversion tended to increase increasing ethanol concentration. The ester conversion increases 1.43% with ethanol : castor oil molar ratio from 6:1 to 9:1. When the ethanol : castor oil molar ratio increases from 9:1 to 19:1, the ester conversion increases 5.8%, but the ester conversion had a small increase when the ethanol : castor oil molar ratio increases from 19:1 to 39:1, only 1.2%. It is shown that an increase of the ethanol : castor oil molar ratio above 19:1 did not produce a significant increase in the ethyl ester conversion.

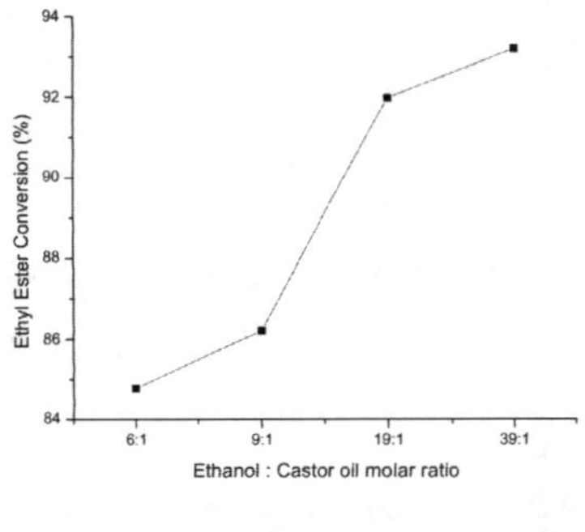

Figure 4 – The effect of ethanol : castor oil molar ratio on the ethyl ester conversion

3.2 Scale up of the reaction

The effect of temperature on ethyl ester conversion was studied for four temperatures. An ethanol : castor oil molar ratio of 19:1 and 1% de sodium hydroxide by weight of castor oil was used. Figure 5 shows that the highest temperature effect was during the first 10 minutes of reaction; after that all reactions achieve the similar ethyl ester conversion. Up to 10 minutes of reaction, the increase of the temperature contributes to an increase of the ethyl ester conversion. The ethyl ester conversions obtained with the scale up had been similar to the ester conversion in laboratory scale.

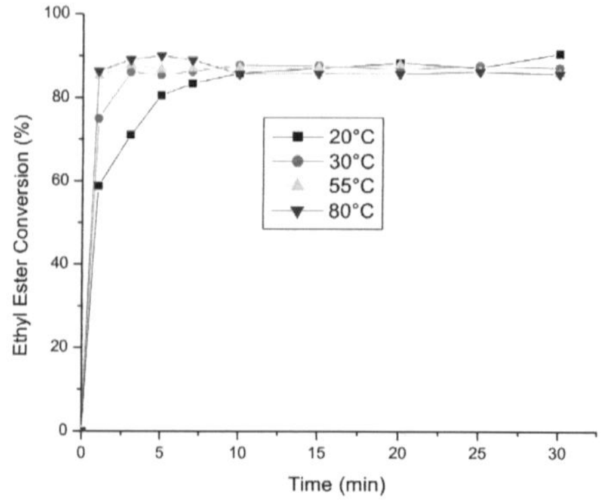

Figure 5 – Effect of temperature on the ethyl ester conversion

4.2. Discussion and Conclusion

The simplified RSM equation takes into account only the significant coefficients. Using the reduced model (eq.2), the response surfaces and contour curves (Figures 6, 7 and 8) were build up. The three parameters (X_1, X_2 and X_3) have statistically significant effect on the esters conversions. The coded results are obtained through a quadratic equation.

$$Y = 89.87 + 0.28 * X_1 + 12.09 * X_2 - 8.07 * X_2^2 + 3.26 * X_3$$
$$- 1.57 * X_3^2 + 0.43 * X_1 * X_2 - 2.81 * X_1 * X_3 - 4.64 * X_2 * X_3 \quad (2)$$

In this equation, the effects of the coded variables can be compared directly by comparing the coefficients of the equations. For example, the effect of the catalyst concentration (X_2) on the ester conversion (Y) is about 43 times higher than of the temperature (X_1) 12.09 vs 0.28. The esters conversion increases at the higher catalyst concentration, temperature and ethanol: castor oil molar ratio, hence the coefficients are positive (b_1, b_2, b_3).

The quality of the fitted models was evaluated by the analysis of variance (ANOVA), based on F-test [5]. As a practical rule, the regression can be considered useful to predict values when the calculated F value, where $\left(F_{calc} = \frac{Mean\ square\ regression}{Mean\ square\ residual} \right)$ is more than the critical F_{listed} [8].

Table 4 shows the ANOVA for the full quadratic model. The reduced model may predict the castor oil transesterification, because the F_{calc} should be eleven times higher than F_{listed}, at 95% of confidence.

Table 4 - ANOVA for the full quadratic model plus three central points

F_{listed} values are significant at the 95% confidence level.
Regression coeficient R = 0.99;

Source of variation	Sum quadratic	Degrees of freedom	Mean quadratic (MQ)	F_{test}	
				F_{calc}	F_{listed}
Regression	3178.1	8	397.26	39.81	3.44 (F0.95;8;8)
Residual	79.82	8	9.98		
Lack of fit	79.7	6	13.28		
Pure error	0.115	2	0.06		
Total	3157.93	16			

Figure 6 shows how the catalyst concentration and the temperature affect the esters conversion. The response surface indicates that, independently of the reaction temperature, the conversion of ethyl ester increase increasing the catalyst concentration. Maximum ester conversions, above 91.65%, were obtained for catalyst concentration from 1.06 to 1.36 wt%, at lower temperature range. Catalyst concentration is an important factor during transesterification, and its effect is positive. There is a moderate increase in conversion with the temperature because the temperature effect is positive, but smaller than the catalyst concentration effect. At higher temperature, a maximum ester conversion was obtained for catalyst concentration from 1.01 to 1.46 wt%.

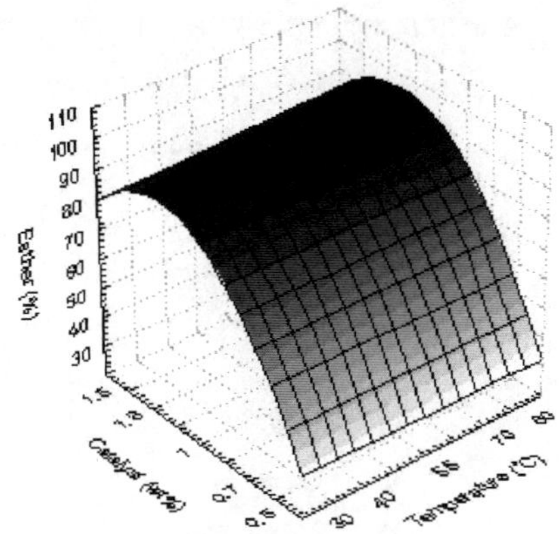

Figure 6 – Response surface and Contour curve of ester conversion vs. catalyst concentration and temperature

In Figure 7, it can be seen how the temperature and the ethanol:castor oil molar ratio affect ester conversions. The castor oil needs a large ethanol excess for the alkaline ethanolysis in order to achieve good conversion. Maximum ester conversions, above 98.6%, were obtained for ethanol:castor oil molar ratio upper to 19:1, within the lower temperature range (30-35°C).

At lower ethanol:castor oil molar ratio, there is a moderate increase in conversion with temperature (94.12%), because the temperature effect is positive, but smaller than that ethanol:castor oil molar ratio.

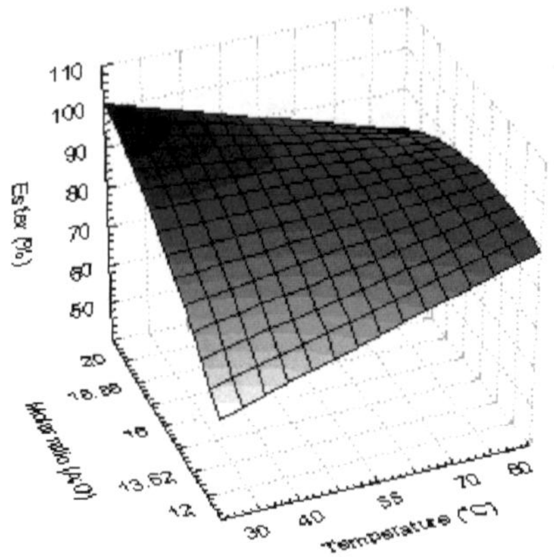

Figure 7– Response surface and Contour curve of Ester conversion vs Molar ratio and Temperature

Figure 8 shows how the catalyst concentration and ethanol:castor oil molar ratio affect ester conversions. The response surface indicates that the catalyst concentration is the most significant factor. Maximum ester conversions, above 88.61%, were obtained for catalyst concentration upper to 1.15wt% at low ethanol:castor oil molar ratios or for catalyst concentration from 0.83 to 1.27wt% at higher ethanol:castor oil molar ratios. The ester conversion has a moderate increase with ethanol:castor oil molar ratio, because this effect is positive and smaller than the catalyst effect.

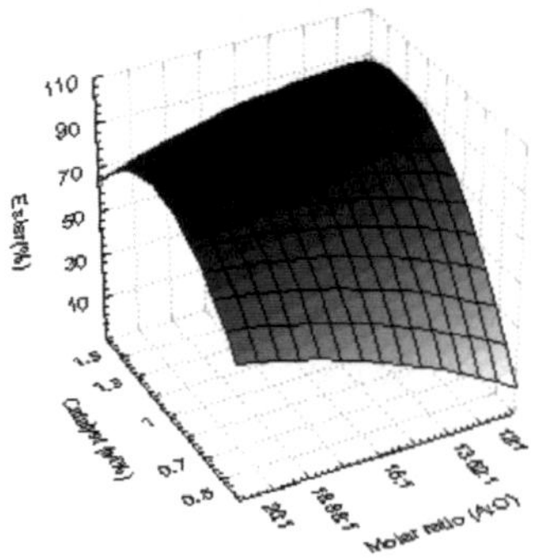

Figure 8 - Response surface of Ester conversion vs. catalyst concentration and Molar ratio

4.3. Concluding Remarks

The aim of this study was to evaluate the different variables affecting the alkaline ethanolysis of castor oil. A second-order model was obtained to predict conversions as a function of temperature, ethanol:castor oil molar ratio and catalyst concentration. Among the 3 parameters studied, catalyst concentration is the most important transesterification factor. The 3 parameters increase the ester conversion, because they have positive influence on the response. Higher ethyl ester conversions are obtained at 30°C, with large catalyst concentration (1.0 – 1.36 wt%) and with large ethanol:castor oil molar ratio, upper to 19:1. So, the statistical model generated may predict the castor oil ethanolysis, because it describes the experimental range adequately. The ethyl ester conversions obtained with the scale up were similar to the ester conversion in laboratory scale.

5 References

[1] Beltrão, N. E. M. – www.tierramerica.net/2003/0526/ianalisis.shtml acess in 29/09/05.

[2] Silva Lima N.*, Wolf Maciel M.R., Batistella C.B., Maciel Filho, R. - Applied Biochemistry and Biotechnology, vol. 129-132, 2006.

[3] Schoenfelder W. –Eur. J. Lipid Sci. Technol. 105, 45-48, 2003.

[4] Filléres R., Benjelloum-Mlayah B. e Delmas M(1995). –J. Am. Oil Chem. Soc. 72 (427-432).

[5] Box, G. E. P., Hunter, W. G. and Hunter J. S.(1978) – Statistics for Experimenters. An Introduction to Design, Data Analysis and Model Building, Wiley, New York.

[6] Mason, R. L., Gunst, R. F., and Hess, J. L.(1989) Statistic Design and Analysis of Experiments: with Applications to Engineering and Science, 2nd edn, John Wiley & Sons, New York.

[7] Kamimura E. S., Medieta O., Rodrigues M. I. and Maugeri F - Biotechnol. Appl. Biochem 33; 2001.

[8] Barros Neto, B., Scarminio, I. S, Bruns, R. E.(2003) – Como fazer Experimentos. Editora Unicamp.

[9] Vicente, G., Coteron, A., Martinez, M.,Araci, J.(1997)– Industrial Crops and Products 8 (29-35).

[10] Freedman, B., Pryde, E.H., Mounts, T.L - J. Am. Oil Chem. Soc. 61,1638-1643, 1984.

[11] Hénon, G., Vigneron, P. Y., Stoclin, B., Calgnez, J. – Eur. J. Lipid Sci. Tecnol. 103, 467 – 477, 2001.

Selective CO oxidation in the presence of excess hydrogen over promoted Pt catalyst

Akkarat Wongkaew, Erdogan Gulari*[a]

Chemical Engineering Department, Faculty of Engineering, Burapha University, Bangsaen, Thailand 20131
[a] *Department of Chemical Engineering, College of Engineering, University of Michigan, Ann Arbor, MI U.S.A. 48109*

The poisoning of a PEM fuel cell electrode by a trace amount of CO contaminating in fuel gases lowers the performance of PEM fuel cells. To eliminate this amount of CO, we can use the method of selective CO oxidation by adding oxygen into gas stream containing the suitable catalyst. The amount of CO contamination should be reduced to less than 10 ppm before the fuel gas enters the PEM fuel cell. This catalyst should be selectively active to CO oxidation reaction but not active to H_2 oxidation reaction. In this work, we developed platinum based catalyst by promoted it with ceria which has an oxygen storage property. We prepared our catalyst by sol gel method. Then, we tested the activity of a 5%Pt/15%CeO_2/Al_2O_3 to selective CO oxidation with the simulated reformed gas composition (1%CO, 20%CO_2, 60%H_2, 10%H_2O and balance with He). We studied the effects of H_2O and CO_2 to the activity of the catalyst. We found that the presence of CeO_2 in a platinum based catalyst dramatically enhanced the performance of the catalyst to CO oxidation. Water has a positive effect to the performance of this catalyst to CO oxidation. CO nearly completely converted to CO_2 at 150°C with the selectivity of 52%. On the other hand, the presence of 20%CO_2 in the gas stream has a negative effect to the activity of the catalyst. This caused by the reverse water gas shift reaction and the formation of carbonate species over ceria sites.

Keywords: selective CO oxidation, platinum, ceria, sol-gel, PEM fuel cells

INTRODUCTION

Steam reforming of methanol [1] or partial oxidation of hydrocarbons [2] followed by water gas shift reaction [3] are commonly used methods to produce H_2 for polymer electrolyte fuel cells (PEFC). The gas composition from these processes contains 40-75%H_2, 15-20%CO_2, ~10%H_2O, 0-25%N_2 and 0.5 to 1%CO. CO contained in the gas stream seriously depresses the performance of PEM fuel cell by poisoning the Pt-based anode electrode of PEM fuel cells [4,5]. Palladium-based membrane purification [6], methanation [7] and selective catalytic CO oxidation [8] are the methods of reducing CO in the gas stream to 10 ppm or less, necessary for operation of the fuel cell, with minimum loss of hydrogen. Membrane purification requires high temperature and high pressure to remove very little concentration of CO in the reformed gas. Methanation requires two moles of hydrogen per mole of CO: $CO+2H_2 \rightarrow CH_4$. This causes not only the reduction of fuel but also the formation of CH_4 to be known as a green house gas. Of these three methods selective CO oxidation is the most promising and economical approach. The selective catalytic CO oxidation process requires a suitable catalyst to have CO oxidation occurred with less or without H_2 oxidation. Many researchers have been searching and developing catalysts for this particular reaction. Platinum over different supports have intensively been studied and shown the promising performance to this application [9, 12].

This brought about the objectives of this work. We developed platinum over alumina catalyst by adding the small amount of cerium oxide into the alumina support. We chose sol gel method for support preparation because we believed that sol gel method offers uniformity and high specific surface area to the final oxide. We selected cerium oxide because of its capacity as an oxygen storage compound. It has been used in many applications such as for three way catalysts under the condition of the lack of oxygen [13] and also for water gas shift reaction enhancing the CO removing from reformed gas to be more fuel (H_2).

EXPERIMENTAL

A. Catalyst Preparation

Cerium aluminum oxide supported Platinum (Pt) catalysts with 5.0 wt% Pt loading was prepared by sol gel technique [14] and incipient wetness impregnation. The desired amount of aluminum isopropoxide was weighted, and then added into a beaker of hot deionized water at 80 °C. After 30 min of aging with continuous stirring, nitric acid (HNO_3) was added to start the hydrolysis reaction resulting in a fibrillar sol. Next, the known amount of cerium nitrate powder was incorporated into the solution. The solution was stirred overnight. Then the obtained solution was kept at 60°C until it became gel. The gel was dried in air at 110°C overnight, and then calcined at 500°C for 13 hours. After calcination, the obtained solid oxide was ground and sieved to a 100-140 mesh powder. For deposition of platinum into the support, we used an incipient wetness impregnation. This method, we added a desired amount of solution of hydrogen hexachloroplatinate (IV) hydrate into cerium aluminium oxide support and then mixed them together until the mixture was uniform. Then the obtained solid was dried in air at 110°C overnight, and then calcined at 500°C for 13 hours. Before testing the activity of this catalyst to selective CO oxidation under each condition, the catalyst was purged with H_2 for 4 hrs.

X-ray diffraction was used to identify the phases present in our samples. A Rigaku Rotating anode X-ray diffractrometer system generating CuKα radiation was used to obtain XRD patterns. The BET surface area and average pore radius of catalysts were measured by an Autosorb-1 Gas Sorption System (Quantachrome Corporation).

B. Activity testing

The catalyst was tested in an integral flow reactor consisting of a flow controller manifold and temperature controller. The K-type thermocouple was placed on the top of the catalyst bed to measure and control the reaction temperatures. The amount of catalyst used was 68 mg and the total gas flow rate was 80 cc./min. An average space velocity was 70,000 cc/g/hr. The inlet gas and outlet gas were analyzed their composition with a gas chromatograph and occasionally with FTIR to determine the limiting concentration of CO.

RESULTS AND DISCUSSION

A. Activity of a 5%Pt/15%CeO_2/Al_2O_3 to selective CO oxidation reaction

We tested the 5%Pt/15%CeO_2/Al_2O_3 under the simulated reformed gas with the addition of 1%O_2 as the result shown in figure 1. From figure 1, the gas composition used was 1%CO, 1%O_2, 10%H_2O, 20%CO_2, 60%H_2 and the rest of He. We observed an increasing of CO conversion from 58% to 97% with an increase of reaction temperatures from 90°C until 170°C. However, selectivity decreased with an increase of reaction temperatures. CO completely converted to CO_2 at 180°C to 200°C. After this temperature, we further increased reaction temperatures up to 230°C and we observed a decrease of CO conversion to 85%. This caused by the desorption of CO at temperatures greater than 160°C leaving Pt active sites for other gas molecules and this led to the competition of CO oxidation, H_2 oxidation and the reverse water gas shift reaction as being confirmed by selectivity. We defined the selectivity of CO oxidation from the ratio of the amount of O_2 used for CO oxidation reaction to the total amount of O_2 consumed by the reactions. We also tested the activity of this catalyst to reverse water gas shift reaction with a gas composition of 25%CO_2 and 75%H_2. We found that at temperature less than 120°C, no CO was observed. However, at reaction temperature of 140°C, CO started forming with less concentration ~ 230 ppm. We increased reaction temperatures to 170°C, 200°C and 230°C. We observed and increase of CO formation from 830 ppm., 25,300 ppm., and 68,300 ppm., respectively.

As can be seen from figure 1, we believed that the 5%Pt/15%CeO_2/Al_2O_3 is one of promising catalysts to selective CO oxidation under the presence of excess H_2 because it works well under the real gas composition with 100%CO conversion, 52% selectivity and the reaction temperature of 190°C.

B. Effect of CO_2

The catalyst was tested under 1%CO, 1%O_2, 20%CO_2, 60%H_2 and balance with He. We cut H_2O from the gas feed in order to study the effect of CO_2 to selective CO oxidation and the results were shown in figure 2. We found that the light off peak was shifted to the right. This means the presence of 20%CO_2 has a negative effect to CO oxidation in the presence of excess H_2. CO completely converted to CO_2 at temperature ~190°C. This may cause from the adsorption of CO_2 over alumina and cerium oxide and this in turns blocking the process of O atom in cerium oxide lattice transferring to CO adsorbed over Pt at the interface between Pt and cerium oxide. However, at temperature above 180°C, CO_2 started desorbing from cerium oxide surfaces and this may cause the completion of CO oxidation at temperatures above 180°C. However, at temperature greater than 200°C, we observed CO peak. Again, this caused by the competition between three reactions: CO oxidation, H_2 oxidation and reverse water gas shift reactions as mentioned above.

C. Effect of H_2O

The catalyst was tested under 1%CO, 1%O_2, 10%H_2O, 60%H_2 and He as balance. The result was shown in figure 3. As can be seen from figure 3, an increase of reaction temperatures from 90°C to 150° increased CO conversion from 60% to 98% (nearly complete conversion). We observed complete CO conversion at the temperature ranges of 160°C to 190°C with selectivity of ~50-54%. Further increasing of reaction temperatures caused less CO oxidation. This incident was happened the same as above cases that at high temperatures, the competitions between CO oxidation, H_2 oxidation and reverse water gas shift reactions were the major cause of the CO formation. Indeed, the presence of H_2O without CO_2 in the gas feed dramatically increases the activity of the catalyst.

CONCLUSIONS

The 5%Pt/15%CeO_2/Al_2O_3 has shown an incredible performance to selective CO oxidation in the excess of H_2 even under the real gas composition. We also found that CO_2 has a negative effect to CO oxidation under this condition. CO completely converted to CO_2 at temperature ~190°C. This may cause by the reverse water gas shift reaction ($CO_2+H_2 \Leftrightarrow CO+H_2O$) and lower the concentration of fuel gas and in turn lower the efficiency of fuel cells. However, when we removed CO_2 from the gas stream and added with 10%H_2O, we observed that water showed a positive effect to selective CO oxidation and we obtained complete CO conversion at low temperatures (~150°C). Temperatures shifted down by 40°C.

ACKNOWLEDGEMENT

Financial support of this research by TRF Thailand is gratefully acknowledged: TRF Grant # MRG 4680140. Also we would like to thank Prof. Levi Thompson and his graduate students for helping me with catalyst characterization and for allowing me to use some instrument in his laboratory.

REFERENCES

1. Amphlett, J.C., Creber, K.A.M., Davis, J.M., Mann, R.F., Peppley, B.A., and Stokes, D.M., Int. J. Hydrogen Energy, 1994, 19, 131.
2. Houseman, J., and Cerini, D.J., SAE Paper No. 740600, 1974.
3. Bunluesin, T., Gorte, R.J., and Graham, G.W., Appl. Catal. B, 1998, 15, 107.
4. Schmidt, V.M., Oetjen, H.-F., and Divisek, J., J. Electrochem. Soc. 144, L237, 1997.
5. Lemons, R.A., J. Power Sources, 1990, 29, 251.

6. Uemiya, S., Kude, Y., Sugino, K., Sato, N., Matsuda, T., Kikuchi, E., Chem Lett., 1988, 10, 1687.
7. Poirier, M.G., Perreault, C., Couture, L., and Sapundzhiev, C., Proc. 1st Int. Sym. On New Materials Fuel Cell System, Montreal, July 1995, p 258.
8. Cohn, A. J.G., US Patent 3,216,783 (November 9, 1965).
9. Manasilp, A., and Gulari, E., Applied Catalysis B: Environmental, 2001, 37, 17.
10. Luengnaruemitchai, A., Osuwan, S., and Gulari, E., International Journal of Hydrogen Energy, 2004, 29, 429.
11. Marino, F., Descorme, C., and Duprez, D., Applied Catalysis B: Environmental, 2004, 54, 59.
12. Liu, X., Korotkikh, O., and Farrauto, R., Applied Catalysis A: General, 2002, 226, 293.
13. Holmgren, A., Duprez, D., and Andersson, B., Journal of Catalysis, 1999, 182, 441.
14. Yoldas, G.E., Ceramic Bulletin, 1975, 54, 298.

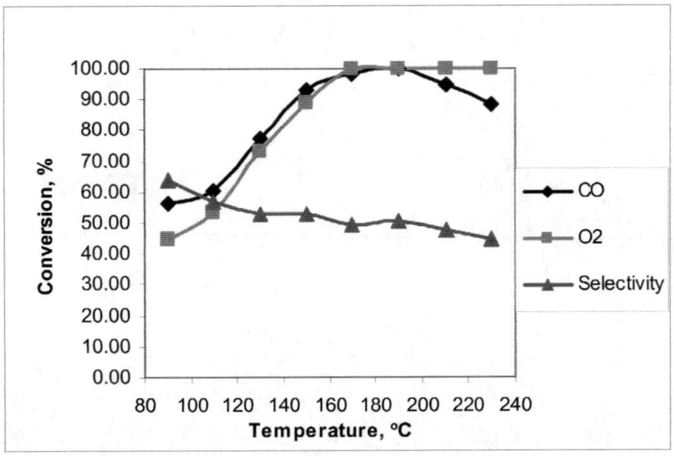

Figure 1 Selective CO oxidation over the 5%Pt/15%CeO_2/Al_2O_3. Gas composition: 1%CO, 1%O_2, 10%H_2O, 20%CO_2 and balance with He. SV=70,000 cc/g/hr.

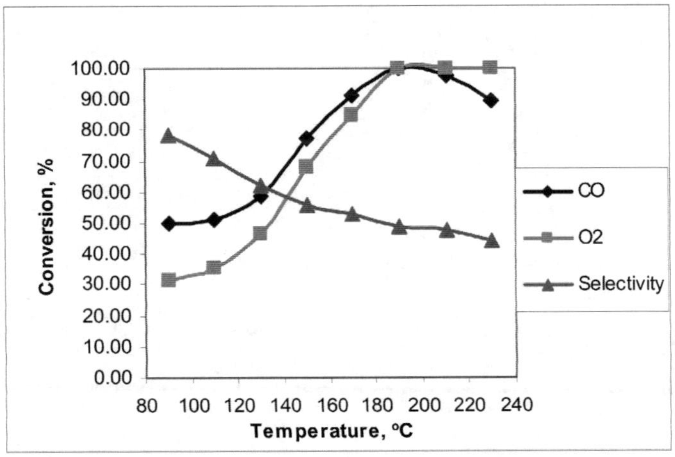

Figure 2 Selective CO oxidation over the 5%Pt/15%CeO_2/Al_2O_3. Gas composition: 1%CO, 1%O_2, 20%CO_2 and balance with He. SV=70,000 cc/g/hr.

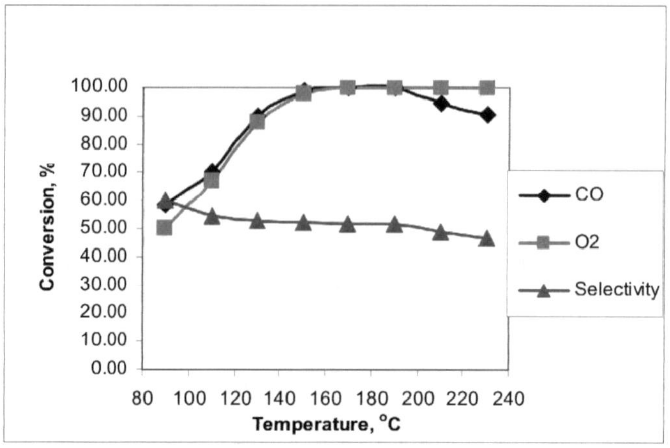

Figure 3 Selective CO oxidation over the 5%Pt/15%CeO_2/Al_2O_3. Gas composition: 1%CO, 1%O_2, 10%H_2O and balance with He. SV=70,000 cc/g/hr.

63d Selective Catalytic Oxidation of Co in the Presence of Hydrogen over Au/Mno2-Ceo2 Catalysts

Hsin-yu Lin, Department of Materials Science and Engineering, National Dong Hua University, 1, Sec. 2,Da Hsueh Rd., Shou-Feng, Hualien, Taiwan and **Yu-Wen Chen**, Department of Chemical and Materials Engineeing, National Central University, 300 Chung-Da Rd., National Central University, Chung-Li, Taiwan.

Supported gold catalyst has been a subject of intense investigation since the report of its exceptionally high CO oxidation activity at low-temperature and high selective oxidation of CO in hydrogen stream (PROX). For fuel cell system,proton exchange membrane fuel cells (PEMFC), a critical issue is to feed carbon monoxide free hydrogen to the anode. Therefore, removal of CO in H2 stream is an important subject in fuel cell performance. The catalyst should have the high activity to oxidize CO without oxidizing H2.It has been reported that Au/CeO2 catalyst has high CO conversion but low selectivity to oxidation of CO, and Au/MnO2 has low CO conversion but high selectivity to oxidize CO.The aim of this study was to investigate the effect of adding MnO2 in Au/CeO2 catalyst on PROX reaction,and to develop a catalyst which has the advantages of single oxide-supported catalyst. In this study, nanosized gold supported on CeO2, MnO2 and MnO2-CeO2 binary oxide were prepared by deposition-precipitation method. The catalytic properties of Au/MnO2-CeO2 were investigated for CO oxidation in H2 stream. The catalysts were characterized by powder X-ray diffraction, N2 sorption, transmission electron microscopy, and X-ray photoelectron spectroscopy analysis. The results showed that gold particles prepared in this study were less than 4 nm. XPS showed that gold were in electron-deficient state. The presence of MnO2 increased the concentration of metallic gold species. The results of PROX reaction showed that the CO conversion of all samples increased with increasing reaction temperature ,whereas the selectivity to CO oxidation decreased with the increasing reaction temperature. For gold on single oxide-support, Au/CeO2 exhibited a much higher CO conversion than Au/MnO2. However, the selectivity of Au/CeO2 decreased with reaction temperature significantly than Au/MnO2. By regulating the preparation and pretreatment procedure and the compositions of Au/MnO2-CeO2 catalyst, we have developed a Au/MnO2-CeO2 catalyst demonstrating excellent performance on both CO conversion and selectivity to CO oxidation at 353 K. This catalyst was also very stable after 3000 h time on stream

Neural Modeling of a Continuous Alcoholic Fermentation Process and Its Optimization by Successive Quadratic Programming

Elmer Ccopa Rivera, Aline C. da Costa and Rubens Maciel Filho
DPQ/FEQ/UNICAMP, Campinas, SP, Brasil – Cx. Postal 6066, 13081-970.
Phone number: +55-19-37883970; Fax: +55-19-37883910.

Abstract

This work focuses on the process operation aspects using model-based optimization of an continuous alcoholic fermentation process. A data-driven identification method based on Multilayer Perceptron Neural Network (MLPNN) and optimal design of experiments was development. The neural model is optimized using Successive Quadratic Programming (SQP) to find out the optimal operational conditions so that the conversion is maximized to the defined allowable limits. In order to check the validity of the computational modeling, the results were compared to the optimization of a deterministic model, whose kinetic parameters were experimentally determined. It was observed that the values for productivity and conversion obtained using the MLPNN models are similar to that obtained using the deterministic model.

Keywords: Continuous alcoholic fermentation process; Optimization; Artificial Neural Network; Successive Quadratic Programming

1. Introduction

Bioethanol (ethanol from biomass) is an attractive, sustainable energy source to fuel transportation. Based on the premise that fuel bioethanol can contribute to a cleaner environment and with the implementation of environmental protection laws in many countries, demand for this fuel is increasing (Zaldivar et al., 2001).

Changes in the operational conditions are quite common in plants of alcoholic fermentation; they occur not only due to the changes in the quality of the raw material but also due to variations of dominant yeast in the process. One of the sources of raw material quality fluctuations is the alteration in the amounts of treated cane juice and molasses used in the composition of the medium fed to the fermentor. These alterations are made due to the sugar factories operation. Furthermore, molasses undergoes variation of composition in different crops. On the other hand, variations in the yeast population are frequent due to the high microbial load in the raw material.

The lack of robustness can be corrected by adjustments in the operational and control parameters of the process when fluctuations occur. In order to accomplish this, it is important that a mathematical model is available to aid in the decision making, mainly when the difficulties of monitoring the key process variables (concentrations of biomass, substrate and product, and temperature) are taken into account. However, the operational changes described make the prediction of the dynamic behavior of the process with a single model difficult, as they lead to changes in microorganism kinetics. Thus, it would be of great advantage to have a mathematical model that could be easily adapted to changes in operational conditions.

This work focuses on the process operation aspects using model-based optimization of an extractive continuous alcoholic fermentation process. Silva et al. (1999) have shown that a scheme combining a fermentor with a vacuum flash vessel presents several positive features and better performance than a conventional industrial process (Andrietta and Maugeri, 1994). The objective in optimizing this process is to maximize productivity, although maintaining a high conversion of substrates to ethanol in the fermentor. A comparison is made between the performances of two models when the process is optimized using Successive Quadratic Programming (SQP). The first model used is a deterministic model, whose kinetics parameters were experimentally determined as functions of the temperature, the second is a nonlinear model represented by a Multilayer Perceptron Neural Network (MLPNN). The methodology based on the hybrid algorithm (MLPNN-SQP) was used to determine the best conditions from the interactive effect of four variables: inlet substrate concentration, cells recycle rate, residence time, and flash recycle.

2. Methods

2.1. A neural network mathematical model for the continuous alcoholic fermentation process

High concentrations of ethanol inhibit the fermentation process, particularly when a fermentative medium with high concentration of substrate is used, as is the case in the majority of the industrial processes. Taking these into account, Silva et al. (1999) studied a process of fermentation combined with a vacuum flash vessel, which selectively extracts ethanol from the medium as soon as it is produced.

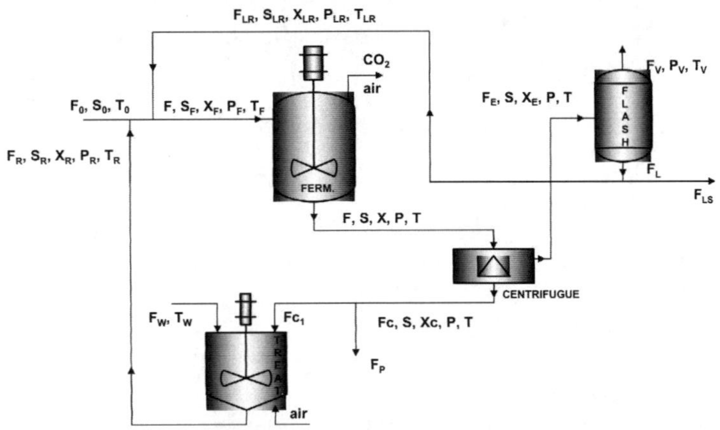

Figure 1. Extractive alcoholic fermentation scheme.

The process to be optimized is shown in Fig.1. The process consists of four interlinked units: the fermentor (ethanol production unit), the centrifuge (cell separation unit), the cell treatment unit, and the vacuum flash vessel (ethanol–water separation unit). The mathematical model is built up of five ordinary differential equations derived from mass and energy balances on the fermentor; algebraic equations for the mass balance on the centrifuge, cell treatment unit, and the purge, as well as the mass and energy balances on the flash tank. Details of the model can be found in Costa et al. (2001).

In order to obtain simulation data for the training of the neural network, the inputs (S_0, R, t_r and r) are distributed according to a design of experiment within the operational intervals (80, 280), (0.2, 0.5), (1.0, 2.5), and (0.2, 0.6), respectively. These intervals were defined based on prior knowledge of the process (Costa et al., 2001). A detailed description of the design of experiments theory can be found in Montgomery (1997). In the present work a full factorial design 2^4 + star configuration with a central point was used, Table 1. In order to have a large amount of training data eight different values of ΔS_0, Δt_r, ΔR, and Δr were used, as shown in Table 2, so that eight factorial designs 2^4 + star configuration were simulated.

Table 1 - Coded factor levels and equations to determine the real values for the input variables used in the training of the neural networks

Coded factor level	S_0	t_r	R	r
Level +2	$(+2)\cdot\Delta S_0(i)+180$	$(+2)\cdot\Delta t_r(i)+1.75$	$(+2)\cdot\Delta R(i)+0.35$	$(+2)\cdot\Delta r(i)+0.4$
Level +1	$(+1)\cdot\Delta S_0(i)+180$	$(+1)\cdot\Delta t_r(i)+1.75$	$(+1)\cdot\Delta R(i)+0.35$	$(+1)\cdot\Delta r(i)+0.4$
Central point (0)	180	1.75	0.35	0.4
Level -1	$(-1)\cdot\Delta S_0(i)+180$	$(-1)\cdot\Delta t_r(i)+1.75$	$(-1)\cdot\Delta R(i)+0.35$	$(-1)\cdot\Delta r(i)+0.4$
Level -2	$(-2)\cdot\Delta S_0(i)+180$	$(-2)\cdot\Delta t_r(i)+1.75$	$(-2)\cdot\Delta R(i)+0.35$	$(-2)\cdot\Delta r(i)+0.4$

Table 2 - Values of ΔS_0, Δt_r, ΔR, and Δr to simulate the data set, used to generate Table 1

i	$\Delta S_0(i)$	$\Delta t_r(i)$	$\Delta R(i)$	$\Delta r(i)$
1	50	0.075	0.375	0.1
2	45	0.0675	0.3375	0.09
3	40	0.06	0.3	0.08
4	35	0.0525	0.2625	0.07
5	25	0.0375	0.1875	0.05
6	25	0.0375	0.1875	0.05
7	20	0.03	0.15	0.04
8	10	0.015	0.075	0.02

2.2. Building up the artificial neural network model

In the last years many works (Chang and Hung, 2002; Chen et al., 2004) have been proposed that use Artificial Neural Networks (ANN) as a substitute for deterministic models. The substitution of the deterministic model by an equivalent ANN at the optimization step presents the advantage of high speed processing.

One of the main current problems related to the alcoholic fermentation process is the lack of robustness of the fermentation in the presence of fluctuations in the quality of the raw material, which leads to changes in the kinetic behavior and influences yield, productivity and conversion. These changes make the prediction of the process dynamic behavior with a single mathematical model difficult. In order to take the kinetic changes into account, the kinetic parameters of the model should be reestimated. However, frequent reestimation of these parameters in deterministic models is usually difficult, mainly due non-linearities, great number of parameters and interactions among them. The use of neural network models is a good option in this case, because the changes in the kinetic behavior can be assessed by updating the network weights.

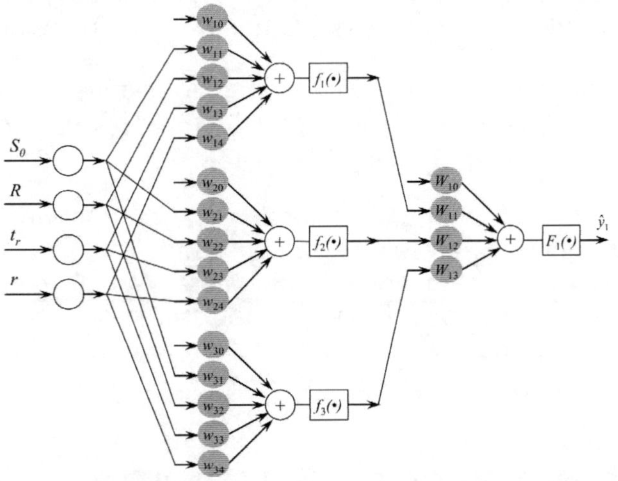

Figure 2 – Representation of a MLPNN with one hidden layer

A MLPNN consists of three types of layers: an input layer, an output layer and one or more hidden layers, as shown in Fig. 2 and described mathematically by Eq. 1. Each layer may have a different number of neurons, and even a different transfer function. An appropriate architecture would help to achieve higher model accuracy. It has been shown that all continuous functions can be approximated to any desired accuracy, with a network of one hidden layer of sigmoidal $f(X) = 1/(1+\exp^{-X})$ hidden units (nodes) and a layer of linear output nodes. Such structure was used in this work.

$$y_1 = F_1[\sum_{j=1}^{M} W_{1j} f_j(\sum_{l=1}^{N} w_{jl} X_l + w_{j0}) + W_{10}] \quad (1)$$

In Eq. (1), $w_{j,l}$ and $W_{1,j}$ specifies the adjustable parameters; i.e., the weights and biases. X is the input variable of the network.

In this work, conversion and productivity were modeled each one with a MLPNN with four inputs (inlet substrate concentration, S_0, cell recycle rate, R, residence time, t_r and flash recycle rate, r) and one hidden layer.

The appropriate number of nodes to include in the hidden layer was addressed with the cross-validation technique in order to avoid model over-fitting. This technique splits the sample data into a training sample and a validation sample. Then MLPNN with different numbers of hidden nodes are trained with the training sample, and their performance monitored with the validation sample in terms of the lowest normalized mean square error (Geva, 1998), given by Eq. 2.

$$NMSE = \frac{1}{\sigma^2 np} \sum_{n=1}^{np} (d_p - x_p)^2 \tag{2}$$

where σ^2 denoted the sample variance of the desired outputs in the test set, x_p and d_p are, respectively, the network and desired outputs and *np* is the number of patterns tested.

In the present work, one half of the data sample was used for training and the other half was intended for validation. The parameters were adjustment using the Levenberg–Marquardt algorithm in Matlab's neural network toolbox.

After the neural networks for conversion and productivity were trained and validated, their parameters (weights and biases) were kept fixed. Thus, the objective of the optimization is to maximize productivity and conversion, which are strongly influenced by the inlet substrate concentration, S_0; cell recycle rate, *R*; residence time, t_r; and flash recycle rate, *r*. In order to accomplish this task, the neural models was written in Fortran to be used by the Successive Quadratic Programming (SQP). The FORTRAN IMSL routine DBCONF was used for this purpose.

2.3. Optimization by successive quadratic programming

Successive Quadratic Programming is used in this work to optimize a given objective function over a given search space. The formulation of objective functions $f(\vec{x})$ is one of the crucial steps in the application of optimization to a practical problem. A general non-linear optimization problem (Biegler and Grossmann, 2004) subjected to the equality and inequality constraints $g_i(\vec{x})$ and $h_j(\vec{x})$, respectively, may be defined as:

$$\text{Optimize } f(\vec{x}) \text{ subject to } g_i(\vec{x}) \le 0,\ i = 1,\ldots,p;\ h_j(\vec{x}) = 0,\ j = p+1,\ldots,m; \\ l_k \le x_k \le u_k,\ k = 1,\ldots,n \tag{3}$$

where $\vec{x}$ is a vector of *n* decision variables $(x_1,\ldots,x_n)$. The l_k and u_k are specified lower and upper bounds on the variables with $l_k \le u_k$.

It is important to properly define constraints on the possible parameter values. First, constraints limit the search space and thus speed up the optimization. Second, physically impossible values for the parameters are avoided, which improves the reliability of the approach. Additional constraints depend on the type of model that is used. In this work the parameters satisfy such conditions as: $80<S_0<280$ kg/m^3, $0.2<R<0.5$, $1.0<t_r<2.5$ h and $0.2<r<0.6$. Costa and Maciel (2004) already used these values successfully in a previous study.

3. Results

3.1. Choice of the neural network architecture

The number of hidden nodes was varied from 2 to 8 and, using the cross-validation criterion to prevent over-fitting, the network with three hidden nodes was found to present the lowest normalized mean square error (Eq. 2) for the validation sample. The results can be seen in Fig. 3.

Productivity and conversion were then modeled both with a MLPNN with 19 scalar parameter

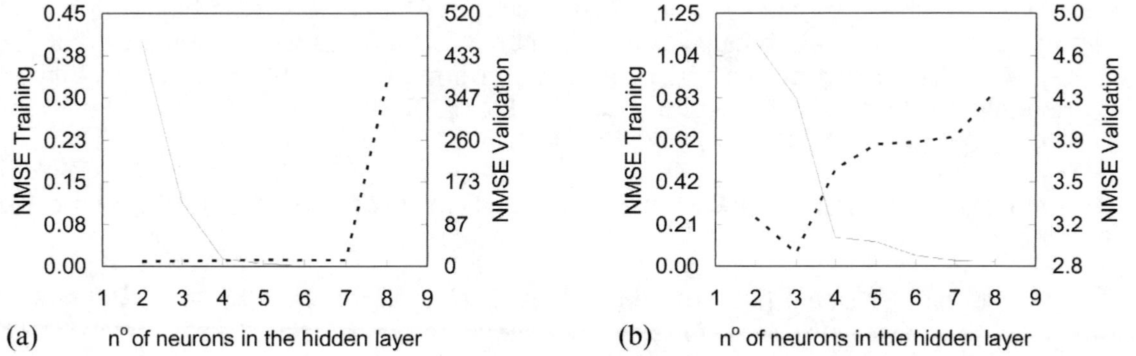

Figure 3 - Training (solid line) and validation (dashed line) error as functions of the number of neurons in the hidden layer: (a) Productivity and (b) conversion

3.2. Optimal solution using successive quadratic programming and deterministic model

The optimization was conducted with the deterministic steady state model of the process. It consists of the steady state mass and energy balances for the fermentor and all the other process units (see Fig. 1). The equations that describe productivity and conversion can be found in Costa et al. (2001).

The optimization problem is to maximize productivity. Thus, the objective function can be formulated as follows:

Maximize *prod* (4)

subject to the equality constraints described by the steady state mass and energy balances for the fermentor and to the inequality constrains:

$conv > 0.99$ (5)
$28 < T < 40°C$ (6)
$80 < S_0 < 280$ kg/m^3 (7)
$0.2 < R < 0.5$ (8)
$1.0 < t_r < 2.5$ h (9)
$0.2 < r < 0.6$ (10)

The optimization variables are: concentration of viable cells, X_v; dead cells, X_d; substrate, S; product, P; and Temperature, T; as well as the variables used by Costa et al. (2001): *S*0, *R*, *tr* e *r.*

One of the main cost contributive parameters for ethanol production from biomass is the cost of the raw material. Such cost can be reduced if the conversion efficiency of the raw material is maximized. Thus, it was included a constraint in conversion (conversion>0.99).

Using the SQP method the following values are obtained: X_v = 47.34 kg/m^3, X_d = 6.42E-9 kg/m^3, S = 1.91 kg/m^3, P = 29.52 kg/m^3, T = 31.28 °C, S_0 = 280 kg/m^3, R = 0.5, tr = 1.0 h and r = 0.55. The values of productivity and conversion were 13.16 kg/(m^3.h) and 0.99, respectively. Although the productivity calculated using SQP technique was higher, the results obtained using this technique are highly dependent on the initial guess. For some values of the initial guess the SQP technique may reach a local maximum or do not converge. In this work the initial guess used was: X_v = 1.0 kg/m^3, X_d = 0.1 kg/m^3, S = 10 kg/m^3, P = 10 kg/m^3, T = 28 °C, S_0 = 161.4 kg/m^3, R = 0.44, tr = 1.6 h and r = 0.37.

3.3. Optimal solution using successive quadratic programming and neural network model

The neural network models can be used as simplified models with the SQP technique. The MLPNN models were structured to explicitly estimate the productivity and conversion from the inlet variables of interest (S_0, R, t_r, and r). This structure models the mass and energy balances for the fermentor and all the other process units. After the solution of the optimization problem, the optimal values of S_0, R, t_r, and r were used in the deterministic model to determined if the MLPNN predictions for optimal productivity and conversion present deviations from the values calculated by the deterministic model.

The optimization problem is postulated as:

Maximize *prod* (11)
$conv > 0.99$ (12)
$28<T<40$°C (13)
$80<S_0<280$ kg/m^3 (14)
$0.2<R<0.5$ (15)
$1.0<t_r<2.5$ h (16)
$0.2<r<0.6$ (17)

The optimized variables values are S_0 = 130 kg/m^3, R = 0.26, tr = 1.07 h and r = 0.59. The values for productivity and conversion calculated by the neural network models are 12.8 kg/(m^3.h) and 0.99, respectively. Using the values calculated for S_0, R, tr and r in the deterministic model it was found productivity of 13 kg/(m^3.h) and conversion of 0.99.

It is possible to notice that, when the optimization problem is solved using successive quadratic programming and the neural network models, the values of temperature and concentrations in the fermentor do not have to be considered as optimization variables and so, the number of the optimization variables is smaller than using deterministic model.

Table 3 shows the optimization results of productivity and conversion calculated by the SQP on the deterministic and MLPNN models. It also shows the comparison among the optimization variables. The results of the optimization methods are expected to lead to different parameters values because they are numerical procedures with expected imprecision in finding the global optimum. Different mathematical models may lead to different predictions, however, a well-trained neural network is able to capture the process dynamic behavior. It is worthwhile mentioning that all the parameters values in Table 3 are within the accepted range for industrial process operation and this was taken into account in the definition of the constraints for the optimization problem.

Table 3 – Optimization variable values obtained through SQP on the MLPNN and deterministic models

	MLPNN-SQP	Deterministic-SQP
S_0 (kg/m^3)	130.0	280.0
R	0.26	0.5
t_r (h)	1.07	1.0
R	0.59	0.55
X_v (kg/m^3)	-	47.34
X_d (kg/m^3)	-	6.42
S (kg/m^3)	-	1.91
P (kg/m^3)	-	29.52
T (°C)	-	31.28
prod (kg/m^3.h)	13.0	13.2
conv	0.99	0.99

4. Discussion

Fermentation processes are quite difficult to model, since their operation involves microbial growth under constantly changing conditions. Hence, there is a need for the development of simple though realistic mathematical models, which can be used to develop optimal operating strategies. The objective is to avoid the decisions based on infrequent measured data generally used in plants of fermentation due to the lack of a robust mathematical model. Under these considerations, the study in this work was focused on the process operation aspects using model-based optimization of an extractive continuous alcoholic fermentation process.

The neural network models was developed using a data set generated from the deterministic model, which means the data are not noisy. However, this may not be considered a limitation, because the procedure may be used when experimental data are available. It can be observed that the values for productivity and conversion obtained using the MLPNN models are similar to that obtained using the deterministic model. MLPNN modeling is a powerful and flexible tool and its use in the alcoholic fermentation process is advantageous because updating the neural network parameters is a simpler procedure than re-estimating kinetic parameters of the deterministic models. Frequent re-estimation of kinetic parameters (or in the case of this work updating the neural network parameters) is necessary owing to changes in the dynamic behavior caused by fluctuations in the quality of the raw material, changes in microorganism metabolism, and variation of the dominant yeast strains present in the fermentation process, among other factors.

In this work, MLPNN models led to a mathematical representation of lower order than the deterministic approach, which intuitively makes the optimization procedure based on hybrid algorithm (MLPNN-SQP) to be significantly quicker. In fact, the neural network model requires only four optimization variables; whereas the deterministic model needs nine.

Experimental design is used to obtain data for the training of the neural network in order to guarantee that the region of interest is covered by the training data. This is a very important issue to be addressed as neural network models have no (or very limited) extrapolation properties.

Although the best result was the one obtained using the deterministic approach, the values for productivity and conversion obtained using the MLPNN-SQP approach are acceptable and can partly eliminate the difficulties of having to specify completely the structure of an alcoholic fermentation model. It is especially promising in situations in which deterministic model is excessively difficult computationally, time-consuming or costly.

5. References

1. Andrietta, S.R. and Maugeri, F., Optimum design of a continuous fermentation unit of an industrial plant for alcohol production. *Advances in Bioprocess Engineering*, 1994, **1** 47-52.
2. Biegler, L., and Grossmann, I. E., Retrospective on optimization. *Computers and Chemical Engineering*, 2004, **28** 1169–1192.
3. Chang, J.S. and Hung, B. C., Optimization of batch polymerization reactors using neural-network rate-function models. *Industrial and Engineering Chemistry Research*, 2002, **41** 2716-2727.
4. Chen, L., Nguang, S.K., Chen, X.D. and Li, X.M., Modelling and optimization of fed-batch fermentation processes using dynamic neural networks and genetic algorithms. *Biochemical Engineering Journal*, 2004, **22** 51-61.
5. Costa, A.C., Atala, D.I.P., Maugeri, F. and Maciel Filho, R., Factorial design and simulation for the optimization and determination of control structures for an extractive alcoholic fermentation. *Process Biochemistry*, 2001, **37** 125-137.
6. Costa, A.C. and Maciel Filho, R., Evaluation of optimization techniques for an extractive alcoholic fermentation process. *Applied Biochemistry and Biotechnology*, 2004 **114/1-3** 485-496.
7. Geva, A.M., ScaleNet-multiscale neural-network architecture for time series prediction. *IEEE Transactions on neural networks*, 1998, 9(5) 1471-1482.
8. Goldemberg, J., Coelho, S.T., Nastari, P.M. and Lucon, O., Ethanol learning curve – the Brazilian experience. *Biomass and Bioenergy*, 2004, **26** 301-304.
9. Montgomery, Douglas C., Design and Analysis of Experiments, 4th ed., John Wiley and Sons, New York, 1997.
10. Silva, F.L.H., Rodrigues, M.I. and Maugeri, F., Dynamic modelling, simulation and optimization of an extractive continuous alcoholic fermentation process. *Journal of Chemical Technology and Biotechnology*, 1999, **74** 176-182.
11. Zaldivar, J., Nielsen, J. and Olson, O., Fuel ethanol production from lignocellulose: a challenge for metabolic engineering and process integration. *Applied Biochemistry and Biotechnology*, 2001, **56**(1-2) 17-34.

64a Performance of a Nonlinear Catalytic Tubular Reactor

Alan Machac, Laurentian University, Ramsey Lake Road, Sudbury, ON P3E 2C6, Canada and **Redhouane Henda**, School of Engineering, Laurentian University, Ramsey Lake Road, Sudbury, ON P3E 2C6, Canada.

Chemical systems in which the kinetics of the reaction is coupled to mass diffusion and convective transport are very common in modern chemical process industries such as bioprocesses, catalysis, and biotechnology. These systems are strongly non-linear due to the nature of the reaction kinetics and internal feedback between the other processes (i.e., convection and diffusion). This non-linearity yields a wide variety of non-trivial behaviours such as multiple steady-states, oscillations, and chaos. Moreover, most chemical processes involve some sort of energy exchange, which in return further adds to the nonlinear nature of the processes and to the palette of complex process behaviour.

Many studies have been conducted with systems consisting of reactions in the presence of diffusion and convection, but most of them have looked at systems involving first order reactions with several assumptions to the system. The corresponding dynamics of the systems were studied while performing several simplifications such as a plug flow adiabatic system, a plug flow adiabatic system with a cooling jacket, a laminar flow adiabatic system with no radial mass transport or axial diffusion, a laminar flow adiabatic system with radial and axial diffusion, and a laminar flow non-adiabatic system.

Mass and energy balance equations can be written for these types of systems in the form of partial differential equations (PDE's). These can then be written in terms of non-dimensional parameters in order to form coupled dimensionless equations which can be solved using a variety of numerical techniques. This study aims at investigating the dynamic behaviour of a two-dimensional catalytic tubular reactor, described via a reaction-diffusion-convection model, using numerical modeling. Different process parameters related to heat exchange in the catalyst bed are assessed and their impact on the behaviour of the system in question is analyzed.

References

Alhumaizi, K., Henda, R. & Soliman, M. (2003). Numerical analysis of a reaction-diffusionconvection system. Computers and Chemical Engineering, 27, pp. 579-594.

Borkink, J.G.H. & Westerterp, K.R. (1992). Significance of axial heat dispersion for the description of heat transport in wall-cooled packed beds. Chemical Engineering Technology, 15, pp. 371-384.

Chan, S.H. & Hoang, D.L. (2000). Chemical reactions in the exhaust system of a cold-start engine. Chemical Engineering Technology, 23, pp. 727-730.

Dekker, N.J.J., Hoorn, J.A.A., Stegenga, S., Kapteijn, F. & Moulijn, J.A. (1992). Kinetics of the CO oxidation by O2 and N2O over Cu-Cr/Al2O3. AIChE Journal, 38, pp. 385-396.

Ganesan, P. & Rani, H.P. (2000). On diffusion of chemically reactive species in convective flow along a vertical cylinder. Chemical Engineering and Processing, 39, pp. 93-105.

Henda, R. & Alhumaizi, K. (2002). Spatiotemporal patterns in a two-dimensional reactiondiffusion-convection system: effect of transport parameters. Mathematical and Computer Modelling, 36, pp. 1361-1373.

64b Development of Different Strategies of Refrigeration for Fixed Bed Catalytic Reactor: Mixed Coolant Flow as Optimal Factor

Edvaldo Rodrigo Morais, Igor Ricardo de Souza Victorino, Eduardo Coselli Vasco de Toledo, and **Rubens Maciel Filho**. Department of Chemical Processes - Faculty of Chemical Engineering - Laboratory of Optimization, Design and Advanced Control, State University of Campinas (UNICAMP), Campinas, Brazil

Fixed Bed Catalytic Reactor have served as the workhorse of the chemical and petrochemical industries for many decades, and are still one of the most used gas-solid reactors in industry, in which heat-transfer plays an important role. The relatively simple construction and behaviour of this reactor make is still favored over other types of more advanced and complicated reactors. Therefore is a large incentive for the design of high performance reactors since this may lead to large production rate, almost total yield and selectivity while ensuring safe operation. The design of cooled tubular reactors, involves complex tradeoffs between tube geometry, pressure drop, and heat-transfer area. Thus the behaviour of chemical reactors depends on variations in the inlet conditions, as well as in other physical and chemical parameters of the system. With a suitable process representation it is possible to propose alternative configurations so that high performance can be achieved. This is the scope of this paper, which explores the benefits of both conventional cooling modes, namely co and counter-current, to design a new fixed bed catalytic reactor using both conventional modes of refrigeration simultaneously, so that it is possible to take advantage of the positive features of each cooling mode. As case studies, the catalytic oxidation of the ethanol to acetaldehyde, the catalytic oxidation of benzene to maleic anhydride and the Ethylene oxidation, were considered. These are strongly exothermic reactions, representatives of important classes of industrial processes. The proposal is to evaluate the potential of different cooling modes to obtain temperatures profiles with lower hot spot and minimum difference between the inlet and outlet reactor temperatures, allowing to keep high conversions and safe operation.

Coolant Configurations This work considers an alternate configuration for the circulation of the coolant fluid combining both concurrent and countercurrent cooling. The coolant flow is divided in two streams: one flowing co-currently while other flowing counter-currently (Figures 1 and 2). The objective is to evaluate the effect of this alternative design in the reactor performance and to compare it with the conventional cooling approaches.

1)The coolant stream flows in co-current mode in the interval from $z = 0$ to $z = L1$ and the flow is counter-current from $z = L1$ to $z = L$. This configuration is illustrated in Figure 1 and is called "Alternative Configuration 1".

2)The coolant stream flows in counter-current mode in the interval from $z = 0$ to $z = L1$ and the flow is co-current from $z = L1$ to $z = L$. This configuration is described in Figure 2 and is called "Alternative Configuration 2".

It should be noticed that at $z = L1$ the coolant stream changes direction in both schemes. Thus, the position of L1 varies for each analysis with the temperature profiles being obtained for each of such situations in both configurations. These two alternative designs have different effects on the reactor operation according to the selected L1 value. So the L1 parameter should be carefully selected. It should be pointed out that while the proposed Alternative Configuration 2 tends to be less efficient than Alternative Configuration 1, it may be useful for the case of process with side and consecutive reactions. The use of alternative coolant configurations increases the operational flexibility and it makes possible a more effective control of the reactor temperature profile. The proposed configurations take advantage of co-current cooling, near to the reactor entrance, and countercurrent operation towards the reactor exit.

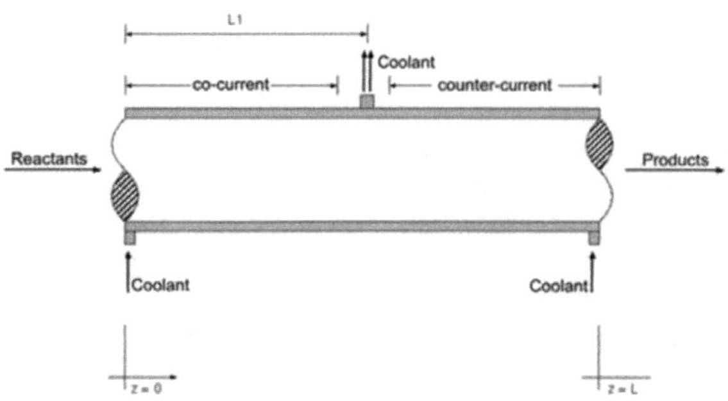

Figure 1. Reactor with Alternative Configuration 1.

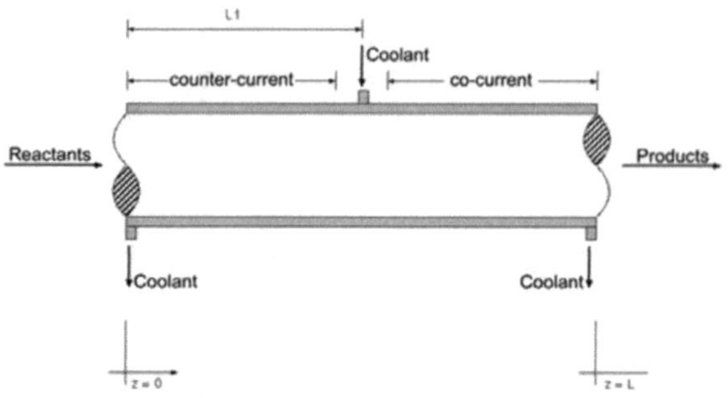

Figure 2. Reactor with Alternative Configuration 2.

64c A Thermal Kinetic Model for the Study of Stereolithography Process

André Luiz Jardini, R. A. Rezende, S.A. Ducatti, R. Maciel Filho, R. A. Rezende, S. A. Ducatti, R. Maciel Filho, R. A. Rezende, S. A. Ducatti, and R. Maciel Filho. DPQ, UNICAMP, Campinas, 6166, Brazil

A numerical and experimental study of thermosetting resins used in infrared stereolithography is presented. In usual practice, stereolithography makes use of photosensitive resins where HeCd (0.352ƒÝm) laser ultraviolet light initiates the curing process. In this work it is considered the process of local curing through the application of infrared radiation, which has proved to be useful in a new technique for the making of prototypes by means of selective heating with a CO2 laser (10.6ƒÝm). The sample consists on a thermosetting resin (epoxy) with the curing agent (diethylenetriamine) and filler (silica). The ideal composition of the thermosetting resins has proved to be 10 parts epoxy, 1.4 part diethylenetriamine (the curing agent) and 0.7 part silica powder. A physical theoretical model is applied to control the parameters which influence the confinement of the curing in the irradiated bulk. A phenomenological model was developed through the solution of the time dependent heat conduction equation in finite elements, which enables the determination of the behavior of curing in terms of irradiation conditions and concentration of reagents. A kinetic model for the study the reaction polymerization processes was developed and describe the main features of the polymerization process, namely diffusion-controlled effects after vitrification and incomplete conversion, presenting a good agreement between experimental and predicted values. An experimental analysis, using Differential Scanning Calorimeter (DSC), was uses to determine the fractional conversion, reaction rate, heat of reaction, and activation energy at which the curing process starts.

64d Melting Heat Transfer in Energy Storage

Subrahmaniyam Susarla Sr., Chemical Engineering, Osmania University, Hyderabad, 500007, India

Freezing and melting are an imporant phenomena in energy storage systems.In this paper,we present melting of two phase change materials(PCM).Octadecane and paraffin wax with melting points of 29 and 60C were used in cylindrical containers with L/d ratios of 1,2 and4.Melting was affected by immersion in isothermal bath with accuracy of 1C.Wall temperatures were maintained 2-25C above the melting point.Data genereated includes melting rate varying with time,temperature distribution in the PCM(radial and axial) and phtographic observations on the interface mavement with time.A critical comparison of literure models with data are presented.

Hestia Distributed Hydrogen Generator

Brook Porter, Kandaswamy DuraiSwamy and Richard Root Woods
Intelligent Energy Inc. 2955 Redondo Ave. Long Beach, California 90806

Intelligent Energy has developed a small scale hydrogen generation appliance, capable of producing 15 kg/day of high purity hydrogen (99.9+) from a range of gaseous or liquid fuels. The device is designed to serve multiple markets, ranging from backup power to vehicle refueling, by producing pure hydrogen from a variety of fuels, including fossil fuels such as FT diesel and renewable fuels such as bio-ethanol.

The Intelligent Energy Hestia hydrogen generation system includes an advanced steam reforming reactor, integrated catalyzed combustion and heat exchangers, compact PSA purification, and low pressure hydrogen storage. The system is capable of reforming gaseous or liquid hydrocarbons to produce 15 kg/day of 99.98+% high purity hydrogen with <1ppm CO. With integrated hydrogen storage, the system is capable of 'instant-on' delivery of pure hydrogen for backup power generation or vehicular refueling.

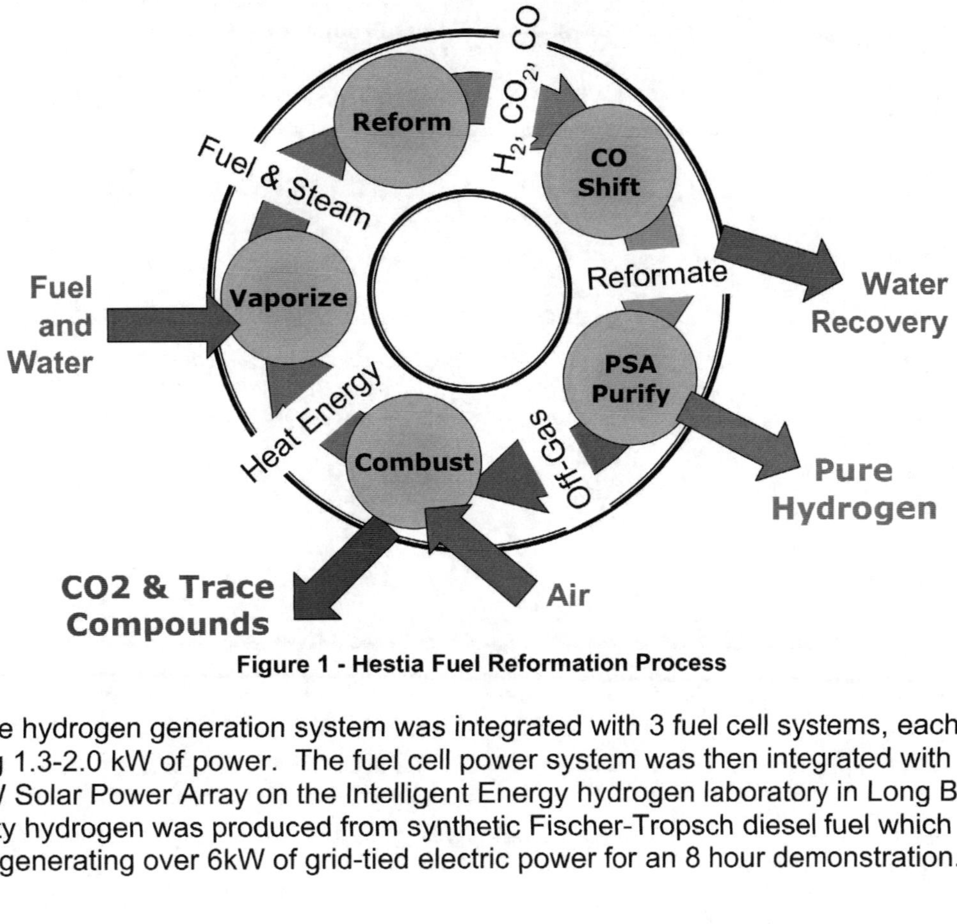

Figure 1 - Hestia Fuel Reformation Process

The hydrogen generation system was integrated with 3 fuel cell systems, each producing 1.3-2.0 kW of power. The fuel cell power system was then integrated with a grid-tied 30kW Solar Power Array on the Intelligent Energy hydrogen laboratory in Long Beach, CA. High purity hydrogen was produced from synthetic Fischer-Tropsch diesel fuel which powered fuel cells generating over 6kW of grid-tied electric power for an 8 hour demonstration.

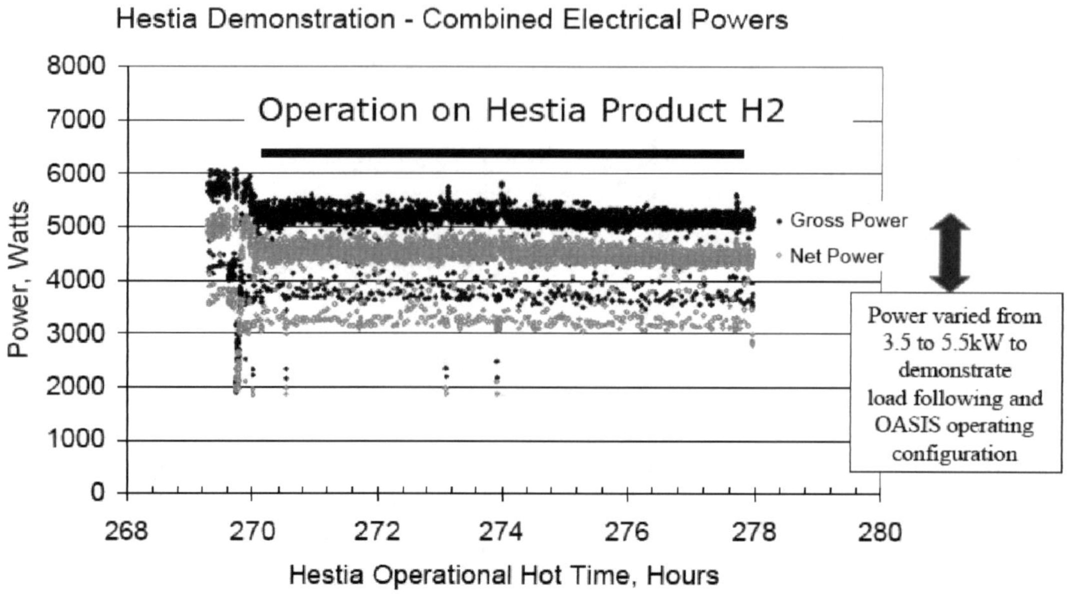

Figure 2 - Electrical Power Production

High purity levels of hydrogen were maintained throughout the 8 hour demonstration, with purity exceeding 99.98% and CO levels below 1ppm.

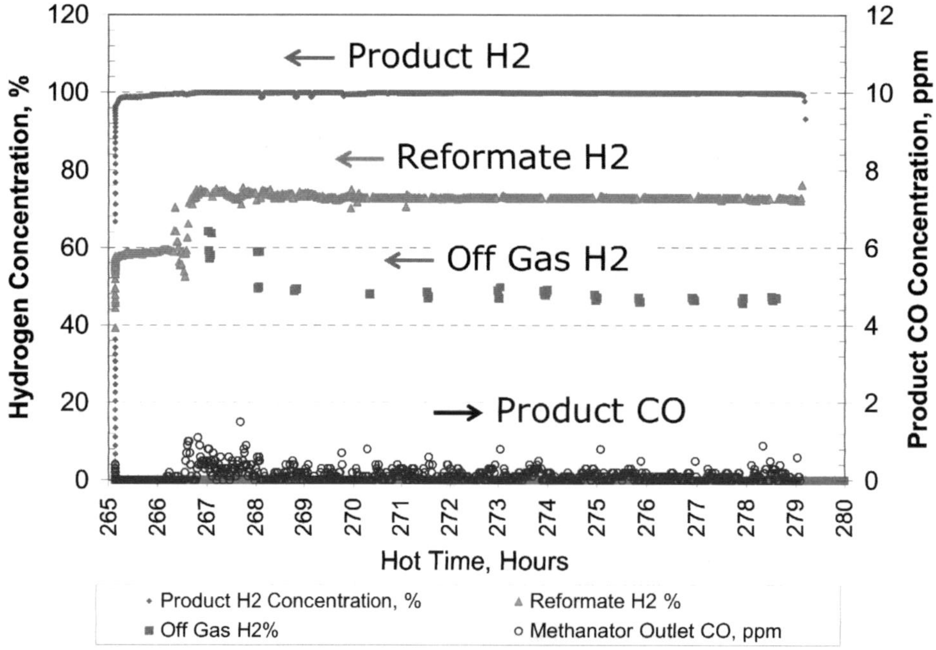

Figure 3 - Stable Hydrogen Purity

With the hydrogen generator and fuel cell system integrated with hydrogen storage, startup time of <10 seconds to full power was demonstrated. Rapid load transients were also demonstrated successfully with the aid of the hydrogen storage buffer.

This demonstration represents one possible configuration for hydrogen generation utilizing renewable solar power and advanced synthetic fuels in a real world application providing useful electrical power and fuel cell vehicle refueling.

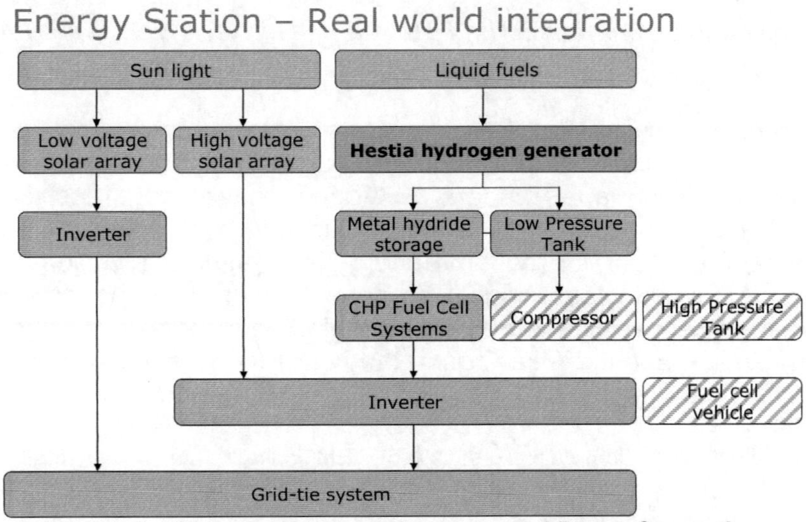

Figure 4 – Open Architecture Hydrogen and Power Generation

Enhanced Waste Energy Recovery Targeting Using Real-Time Energy Integration in Pulp & Paper Plant

M. B. Noureldin[(1)], S. Al-Hashimi[(3)], A.S.Aseeri[(1)], A.H. Qahtani[(1)] and A. K. Hasan[(2)].

[(1)] *Consulting Services Department, Saudi Aramco, Dhahran, Saudi Arabia.*

[(2)]*Dept. of Materials & Process Eng., The University of Waikato, Private Bag 3105, Hamilton, New Zealand*

[(3)] *Chemical Engineering Department, The Petroleum Institute, Abu Dhabi, U.A.E.*

Abstract

Energy conservation has become a major element in controlling operating cost of any industrial facility. Pulp & paper industry is one of the most energy-intensive manufacturing sectors. This paper renders new opportunity for aggressive waste energy recovery in pulp & paper industry through the application of real-time energy integration. Targeting as a feasibility check for real-time energy integration application impact on plant energy consumption is conducted using TEM_icons software. The paper shows that real-time energy integration application in pulp& paper plants for the sake of energy savings is justified and can result in about 8% savings in the hot utility consumption of a typical pulp& paper plant.

Key Words: Real-time Heat Integration; Pulp & paper plant; TEM_icons™; Interval Constraint Satisfaction;

1. Introduction

Pulp and paper industry is one of the most growing sectors in the world. World demand for paper and paperboard is forecasted to grow from the current 300 million tons to over 420 million tons by the year 2010 or an average growth rate of 2.8% per annum [1]. In 2002, North American pulp mills produced 78.2 million tons of paper grade wood pulp. The United States, the largest pulp producer in the world, produced 52.7 million tons in 2002 [2]. Though nearly 60-70% of its energy utilized is derived from wood waste, pulpwood, cogeneration, and a renewable resource, pulp & paper is still an energy intensive industry. This industry accounts for about 5% of total final energy use in International Energy Agency (IEA) member countries, which is about 15% of final use in the industrial sector [3]. In U.S, pulp & paper mills account for about 12% of total manufacturing energy use. The growing demand for energy and the significant increase in its prices have initiated the need for more innovative solutions in energy efficiency optimization.

Many methods and tools for energy conservation have been developed during the last three decades. These methods used mainly mathematical programming and pinch technology to determine the optimal process conditions that result in efficient use of energy in process industries. This paper demonstrates the energy conservation benefits obtained from utilizing real-time energy integration techniques in a typical pulp & paper plant. It uses TEM_icons software to first systematically target for energy savings that may be obtained through the implementation of a real-time energy integration system and second, specifying the optimal process operating conditions that achieve desired target. In industrial community targeting step that can quantify the best and worst case scenarios of energy saving due to the application of real-time integration systems is very important prerequisite engineering study to justify more engineering work towards any new capital investment project.

The application of real-time energy integration in pulp &paper plants is not only beneficial in maximizing energy conservation during plants operation but also in reducing the energy-based air pollution in pulp & paper industry.

2. TEM_icons™ Software at a Glance

TEM_icons™ is interval constraint satisfaction-based software. It is a mass, energy and time management program that presents a new methodology for energy efficiency enhancement systematically. It contains unique capability for optimal energy systems targeting. It is the first user-friendly software that can globally calculate, without exhaustive enumeration, energy utility targets under all possible combinations of process parametric and structural modifications. It can also calculate these targets under all known possible combinations of process disturbances and uncertainties. It can define the optimal process conditions so that optimal utility consumption can be specified. It globally solves the minmin and maxmax energy problems for industrial applications [4]. Solving these problems gives the global minimum and maximum heating and cooling utilities for any plant. While these global energy targets often are not attained together due to the fact that the process conditions that lead to the global minimum in heating utility are not the same process conditions that lead to the global minimum of the cooling utilities, the software determines the targets that can be attained simultaneously [5].

3. Problem Statement

"Given an industrial pulp and paper plant that consists of several processing units, determine systematically (i) the global minimum and maximum energy consumption for both heating and cooling utilities under all possible combination of given process disturbances and (ii) the set of process conditions that if we can properly control would render the global minimum heating utility consumption without exhaustive enumeration".

4. Solution Methodology

Develop a process path model for the whole plant with all possible combinations of process disturbances in a constraint logic propagation environment with interval solver. For a desired production target solve the model to define the flowrate, supply temperature and target temperature intervals of each process stream that need to be cooled or heated [6]. This step captures the effect of process units and streams interactions on the interval values of process streams flow specific heats and supply and target temperatures that need to be cooled or heated due to process disturbances. Since this step is a little time consuming and needs modeling expertise it can be simplified via obtaining these intervals of independent process streams flowrate, supply and target temperatures from experienced process owners [7].

Find best and worst cases scenarios of both heating and cooling utilities consumption under all possible combinations of defined process disturbances. The difference between the best and worst case energy consumption values; is the first indication whether there is an economic incentive to pursue more aggressive waste heat recovery in the plant through real-time heat integration. If the difference between best case scenario of energy consumption and worst case scenario is not worth it, real-time energy integration system as an investment will be most probably not justified. The final step in the targeting study will then be the finding of the set of process conditions that if properly controlled and can be all controlled will achieve the desired energy consumption scenario. This step can be implemented via setting up the desired energy consumption target in the software heating or cooling utility template and screening the interval-based process conditions till we reach the desired energy consumption target. Such procedures are formulated; automatically configured and implemented in the **TEM_icons™** software environment [8,9].

4.1 Process Description Overview

The mill studied in this paper uses mostly virgin fiber to produce high-quality bleached and unbleached kraft market pulp and container board. It has two pulp-mills, one paper machines and one pulp dryer that produce around 600,000 tons of pulp and paper each year as shown in the simplified process flow diagrams in Fig. 1 and Fig. 2.

The process is divided into several stages; wood preparation, pulp-mills, steam generation and recovery section, pulp dryer, recycle fiber plant, and paper machine. The first pulp-mill produces unbleached pulp predominantly for use on paper machine. The mill uses a batch cooking process that pulps softwood chips using the kraft method. Batches of wood chips are mixed with white liquor and cooked at high temperature (170°C) and pressure (870 kPa) in each of the plant's five batch digesters. Then cooked chips are released into the blow tanks at atmospheric pressure. The knots and shives are removed from the pulp by screens to ensure that only clean pulp goes to storage. Pulp is washed to remove cooking chemicals, lignin and other materials in the form of black liquor. Black liquor is recycled to the steam generation and recovery section to be processed for re-use. Pulp is then stored until needed by Paper machine.

The second pulp-mill is able to produce either bleached or unbleached pulps for use on Pulp dryer or on Paper machine. It uses a continuous digester process to pulp chips. It cooks under pressure in white liquor solution.

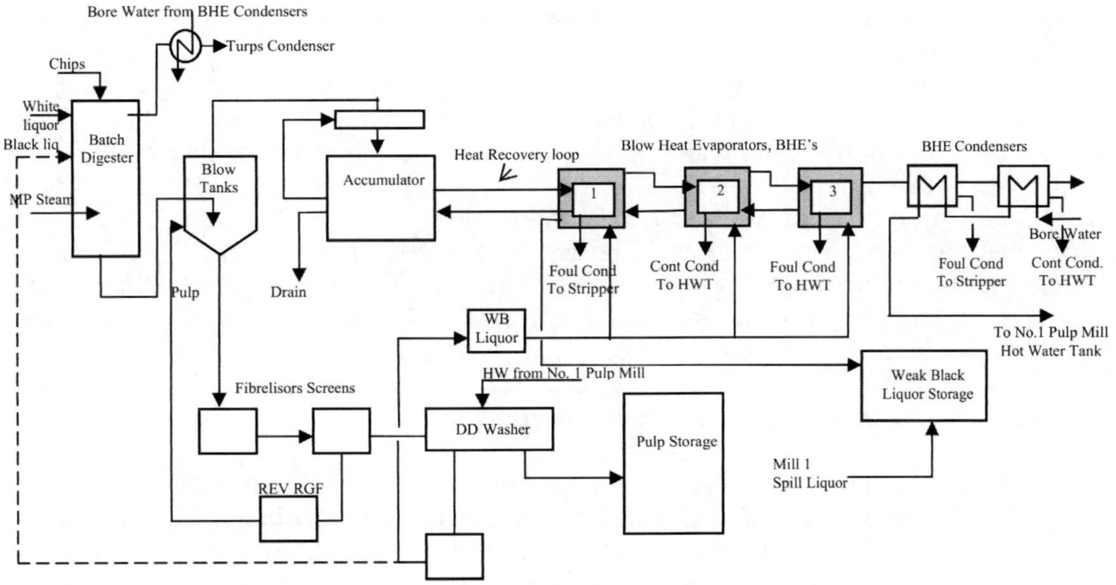

Fig. 1. Simplified process flow diagram for the first pulp mill

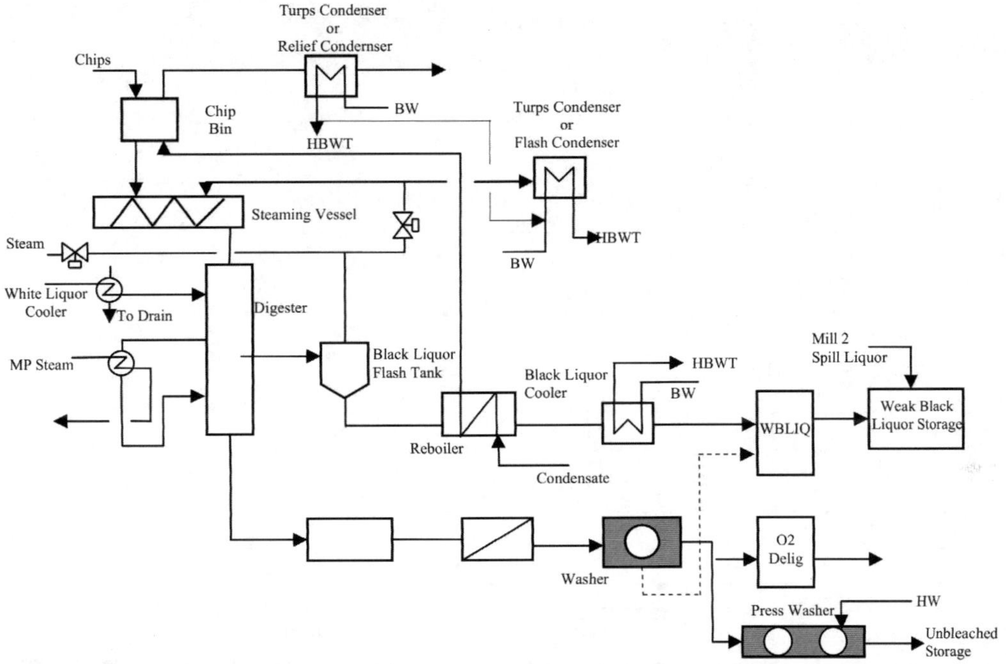

Fig. 2. Simplified process flow diagram for the second pulp mill

Pulp form the continuous digesters are released to blow tank that contains a low-pressure environment where chips blow apart as they travel from the high-pressure area in the digester to the low-pressure area, causing wood fibers to separate. Pulp is screened and washed before going to either a further washing stage before being stored as unbleached pulp or the oxygen delignification plant for pre-processing before going to bleach plant. Pulp which is to be bleached, goes through O2 reactors to provide retention time for steam, oxygen and oxidized white liquor to dissolve lignin. Then press washers remove dissolved lignin and cooking chemicals from pulp.

The recovery section converts weak black liquor into white liquor that can be reused. Black liquor from the pulp-mills consists of large quantities of lignin and also water which enters the liquor during pulp washing processes. Weak black liquor passes through evaporators, which use steam to heat the liquor to drive out water. This process thickens the weak black liquor to produce strong black liquor.

The mill has two boiler systems. The two boilers generate a total of 160 tons/hr. The mill also produces about 50% of the total electricity usage in the co-generation system with a generation capacity of 40 MWatts. The types of fuel used for firing the boiler are natural gas, wood-waste, or a mixture of wood-waste and coal.

4.2 Process Data Extraction for Real-time Energy Analysis

The streams included in our study are the ones allowed by the process owners to be integrated for the sake of minimizing heating and cooling utilities consumption. The pulp &paper plant studied in this paper consists of 22 hot and 27 cold streams. These streams are described in details in Table 1 including the heating and cooling requirements. These streams are the ones in the pulp mills 1 & 2 along with the evaporators and hot water heaters. The hot and cold streams heat duty integration is conducted in this energy study using a minimum temperature difference between the hot and the cold (ΔTmin) of 10°C.

Table1. Data Extraction for Energy Study

Hot Steams					
Stream ID	**Name**	**Inlet Temp(°C)**	**Outlet Temp (°C)**	**FCp (kW/ °C)**	**Enthalpy (kW)**
h1	Blow Accumulator	100	90	1,400.00	14,000.00
h2	Blow Accumulator Drain	80	20	15.00	900.00
h3	Turps Condenser	100	95	720.00	4,000.05
		95	60	11.43	
h4	BHE condenser no 1	80	79	8,000.00	8,000.00
h5	BHE condenser 2	60	59	11,000.00	11,000.00
h6	Drum Clean end Repulp over flow	60	20	121.25	4,850.00
h7	Flash Tank	160	100	257.00	15,420.00
h8	Reboiler	125	110	386.70	5,800.50
h9	Black liquor cooler	110	90	390.00	7,800.00
h10	Relief condenser	91	90	7,500.00	7,500.00
h11	Bleach Plant E Filtrate overflows	70	20	180.00	9,000.00
h12	Bleach Plant Filtrate	60	20	300.00	12,000.00
h13	CD3	101	100	1,800.00	1,800.00
h14	Evap 4 6th Effect a	55	53	3,000.00	6,000.00
h15	Evap 4 6th effect b	55	53	9,400.00	18,800.00
h16	Evap 5 6th effect	55	53	12,200.00	24,400.00
h17	Stripper condenser	80	70	150.00	1,500.00
h18	Vat dilution cooler	90	80	205.00	2,050.00
h19	Filtrate cooler	90	80	290.00	2,900.00

h20	PM 6 Effluents	60	20	208.75	8,350.00
h21	PM 2 Effluents	60	20	690.00	27,600.00
h22	White liquour cooling	80	60	180.00	3,600.00
					197,270.55
		Cold Steams			
Stream ID	**Name**	**Inlet Temp(°C)**	**Outlet Temp (°C)**	**FCp (kW/°C)**	**Enthalpy (kW)**
C1	Chip Heating LP	145	146	5,000.00	5,000.00
C2	Chip Heating MP	183	184	3,000.00	3,000.00
C3	BHE 1st Effect	85	86	14,000.00	14,000.00
C4	Chip Bin heating LP	100	101	18,200.00	18,200.00
C5	Chip Bin heating WH	100	101	5,800.00	5,800.00
C6	Streaming Vessel Heating	110	111	9,000.00	9,000.00
C7	Digester Heating	159	160	15,800.00	15,800.00
C8	O2 Delig Heating	145	146	4,200.00	4,200.00
C9	Hot bore hole water	15	62.34	158.46	7,501.50
C10	Hot bore hole water	32.8	82.8	128.50	6,425.00
C11	Hot bore hole water	15	76.5	126.77	7,796.36
C12	Hot bore hole water	15	40.7	427.85	10,995.75
C13	Hot bore hole water	40.7	79.65	102.70	4,000.17
C14	Hot bore hole water	40.7	71.859	256.74	7,999.61
C15	Hot bore hole water	40.7	67	68.47	1,800.76
C16	Hot bore hole water	73.6	80	791.93	5,068.35
C17	Hot water	15	28	230.77	3,000.01
C18	Hot water	28	40	250.00	3,000.00
C19	Hot water	40	70	68.33	2,049.90
C20	Hot water	40	70	96.67	2,900.10
C21	Hot water	61	80	239.47	4,549.93
C22	Deaerator Feed water heating	15	145	230.77	30,000.10
C23	White Liquor heating	80	170	73.33	6,599.70
C24	Black Liquor heating	80	170	73.33	6,599.70
C25	Evap 4 1st effect	137	138	26,000.00	26,000.00
C26	Evap 5 1st Effect	137	138	26,000.00	26,000.00
C27	Stripper Stream	110	111	1,500.00	1,500.00
					238,786.92

5. Results and Discussions

Current hot utility and cold utility consumption of the pulp and paper plant studied in this paper is152 MW and 111 MW respectively. Heat integration using Pinch analysis conducted nowadays in almost every new industrial facility can result in significant savings in the current utility bill. The new energy targets are 119 MW and 78 MW for heating and cooling utilities respectively. The pinch temperature is found to be at 60°C of the hot streams and 50°C for the cold streams as shown in Fig. 3. The new targets show that both heating and cooling demands can be reduced by about 33 MW.

Active	ID	Flow Name	Type	Temp. In °K	Temp. Out °K	FCP W/(°K)	Enthalpy Min W	Enthalpy Max W
Active	9	Black liquor cooler	Hot	110.00 : 110.00	90.00 : 90.00	390.00 : 390.00	7800.00	7800.00
Active	10	Relief condenser	Hot	91.00 : 91.00	90.00 : 90.00	7500.00 : 7500.00	7500.00	7500.00
Active	11	Bleach plant E Filtrate overfl...	Hot	70.00 : 70.00	20.00 : 20.00	180.00 : 180.00	9000.00	9000.00
Active	12	Bleach plant filtrate	Hot	60.00 : 60.00	20.00 : 20.00	300.00 : 300.00	12000.00	12000.00
Active	13	CD3	Hot	101.00 : 101.00	100.00 : 100.00	1800.00 : 1800.00	1800.00	1800.00
Active	14	Evap 4 6th Effect a	Hot	55.00 : 55.00	53.00 : 53.00	3000.00 : 3000.00	6000.00	6000.00
Active	15	Evap 4 6th Effect b	Hot	55.00 : 55.00	53.00 : 53.00	9400.00 : 9400.00	18800.00	18800.00
Active	16	Evap 5 6th effect	Hot	55.00 : 55.00	53.00 : 53.00	12200.00 : 12200....	24400.00	24400.00
Active	17	Stripper condenser	Hot	80.00 : 80.00	70.00 : 70.00	150.00 : 150.00	1500.00	1500.00
Active	18	Vat dilution cooler	Hot	90.00 : 90.00	80.00 : 80.00	205.00 : 205.00	2050.00	2050.00
Active	19	Filtrate cooler	Hot	90.00 : 90.00	80.00 : 80.00	290.00 : 290.00	2900.00	2900.00
Active	20	PM 6 Effluents	Hot	60.00 : 60.00	20.00 : 20.00	208.75 : 208.75	8350.00	8350.00
Active	21	PM 2 Effluents	Hot	60.00 : 60.00	20.00 : 20.00	690.00 : 690.00	27600.00	27600.00
Active	22	White liquour cooling	Hot	80.00 : 80.00	60.00 : 60.00	180.00 : 180.00	3600.00	3600.00
Active	23	Chip heating LP	Cold	145.00 : 145.00	146.00 : 146.00	5000.00 : 5000.00	5000.00	5000.00
Active	24	Chip heating MP	Cold	183.00 : 183.00	184.00 : 184.00	3000.00 : 3000.00	3000.00	3000.00
Active	25	BHE 1st Effect	Cold	85.00 : 85.00	86.00 : 86.00	14000.00 : 14000....	14000.00	14000.00
Active	26	Chip Bin heating LP	Cold	100.00 : 100.00	101.00 : 101.00	18200.00 : 18200....	18200.00	18200.00
Active	27	Chip bin heating WH	Cold	100.00 : 100.00	101.00 : 101.00	5800.00 : 5800.00	5800.00	5800.00
Active	28	Streaming vassel heating	Cold	110.00 : 110.00	111.00 : 111.00	9000.00 : 9000.00	9000.00	9000.00
Active	29	Digester heating	Cold	159.00 : 159.00	160.00 : 160.00	15800.00 : 15800....	15800.00	15800.00

Network Variables
Interval: 1.00
Delta T: 10.00

Utilities Without Integration
Heating 238787.08 : 238787.08 W
Cooling 197270.55 : 197270.55 W

Utilities With Integration
Heating 119421.01 : 119421.01 W
Cooling 77904.48 : 77904.48 W

ROMC:
Min: 60.00
Max: 60.00

Fig. 3. Energy consumption at plant's nominal operating conditions

To achieve these targets no heat transfer must take place across the pinch temperature which is associated by 15 heat exchangers out of 39. About 86% of the cross pinch heat transfer is linked by 8 heat exchangers. These results mean that a straight forward retrofit project can be implemented to eliminate the inefficiencies by re-arranging and/or establishing new heat exchanger network after identifying the location and size of the individual cross pinch exchangers. These practices became common nowadays in the industrial community. The paper is trying to go beyond this phase via exploring the possibility of installing real-time energy integration system that of-course will need more involved control structure software and hardware modifications for more aggressive waste heat recovery in the process. The pulp and paper plant under given disturbance scheme as we will see in this section exhibits potential for energy saving if we conduct proper control of some streams to keep the process always at its best operating conditions. To target for the saving we used the TEM-icons software to obtain the best case and worst case scenarios of energy consumption under given process disturbances. While in this specific plant application the streams manipulation enjoyed complete dependence, this is not the case in most of industrial application. In our future paper we will explain in details how to use the software to handle streams dependence in energy targeting to avoid over estimation of best case and worst case scenarios plant's energy consumption.

It is instructive to mention here that targeting for utilities under process disturbances are calculated for the given disturbance scheme by the pulp and paper plant process owners. In the software this task of entering process data under given disturbances has been conducted via the fuzzification of the sensible heat streams supply and target temperatures for a given range of disturbance of ±5°C. (process owners suggested that such disturbance scheme of only sensible heat streams supply and target temperatures are the least sever disturbances faced in their plant while also ensuring independence)

Using TEM_icons™ the global heating and cooling utilities consumption targets obtained are 109.4 MW and 62.5 MW for heating and cooling utilities respectively as shown in Figure 4.These results show that the reduction in the heating utility is about 9976.3 kW.

Active	ID	Flow Name	Type	Temp. In °K	Temp. Out °K	FCP W/(°K)	Enthalpy Min W	Enthalpy Max W
Active	15	Evap 4 6th Effect b	Hot	55.00 : 55.00	53.00 : 53.00	9400.00 : 9400.00	18800.00	18800.00
Active	16	Evap 5 6th effect	Hot	55.00 : 55.00	53.00 : 53.00	12200.00 : 12200.00	24400.00	24400.00
Active	17	Stripper condenser	Hot	80.00 : 80.00	70.00 : 70.00	150.00 : 150.00	1500.00	1500.00
Active	18	Vat dilution cooler	Hot	90.00 : 90.00	80.00 : 80.00	205.00 : 205.00	2050.00	2050.00
Active	19	Filtrate cooler	Hot	90.00 : 90.00	80.00 : 80.00	290.00 : 290.00	2900.00	2900.00
Active	20	PM 6 Effluents	Hot	55.00 : 65.00	15.00 : 25.00	208.75 : 208.75	6262.50	10437.50
Active	21	PM 2 Effluents	Hot	55.00 : 65.00	15.00 : 25.00	690.00 : 690.00	20700.00	34500.00
Active	22	White liquour cooling	Hot	75.00 : 85.00	55.00 : 65.00	180.00 : 180.00	1800.00	5400.00
Active	23	Chip heating LP	Cold	145.00 : 145.00	146.00 : 146.00	5000.00 : 5000.00	5000.00	5000.00
Active	24	Chip heating MP	Cold	183.00 : 183.00	184.00 : 184.00	3000.00 : 3000.00	3000.00	3000.00
Active	25	BHE 1st Effect	Cold	85.00 : 85.00	86.00 : 86.00	14000.00 : 14000.00	14000.00	14000.00
Active	26	Chip Bin heating LP	Cold	100.00 : 100.00	101.00 : 101.00	18200.00 : 18200.00	18200.00	18200.00
Active	27	Chip bin heating WH	Cold	100.00 : 100.00	101.00 : 101.00	5800.00 : 5800.00	5800.00	5800.00
Active	28	Streaming vassel heating	Cold	110.00 : 110.00	111.00 : 111.00	9000.00 : 9000.00	9000.00	9000.00
Active	29	Digester heating	Cold	159.00 : 159.00	160.00 : 160.00	15800.00 : 15800.00	15800.00	15800.00
Active	30	O2 delig heating	Cold	145.00 : 145.00	146.00 : 146.00	4200.00 : 4200.00	4200.00	4200.00
Active	31	Hot Bore Hole water- s1-s2	Cold	15.00 : 15.00	62.34 : 62.34	158.46 : 158.46	7501.50	7501.50
Active	32	Hot Bore Hole water- m1-m2	Cold	32.80 : 32.80	82.80 : 82.80	128.50 : 128.50	6425.00	6425.00
Active	33	Hot Bore Hole water-s1-m2	Cold	15.00 : 15.00	76.50 : 76.50	126.77 : 126.77	7796.35	7796.35
Active	34	Hot Bore Hole water-s1-s3	Cold	15.00 : 15.00	40.70 : 40.70	427.85 : 427.85	10995.75	10995.75
Active	35	Hot Bore Hole water-s3-m2	Cold	40.70 : 40.70	79.65 : 79.65	102.70 : 102.70	4000.17	4000.17

Network Variables
Interval: 1.00
Delta T: 10.00

Utilities Without Integration
Heating 235012.78 : 242561.38 W
Cooling 169926.40 : 224614.7(W

Utilities With Integration
Heating 109444.70 : 135210.61 W
Cooling 62575.63 : 99046.62 W

ROMC:
Min: 60.00
Max: 110.00

Fig. 4. Targeting for energy consumption under all possible combination of defined process disturbances

Global minima for the heating and cooling can not be reached with the same set of operating conditions. The results shown in figures 4 and 5 suggest that global heating or cooling minima targets can be reached upon the proper control of specific set of operating conditions. We will need to control only 12 streams out of 49 streams within the given range of disturbance in order to attain such global minimum heating or global minimum cooling utilities. Table 2 shows the optimal operating conditions that render global minima targets. Out of the whole stream list there are two bifurcation streams. They are the operating conditions of streams h7 (Flash tank) and h11 (bleach plant 'E' filtrate overflow) that will decide which global minimum we are going to attain.(global minimum heating utility or the global minimum cooling utility) For the global minimum heating utility target, since it is always more important in pulp and paper industry, the temperature of h7 is set or shall be controlled at 95°C (instead of 100°C) and supply temperature of h11 is set or shall be controlled at 75°C (instead of 70°C). In such case the system pinch point is going to be at 110°C.

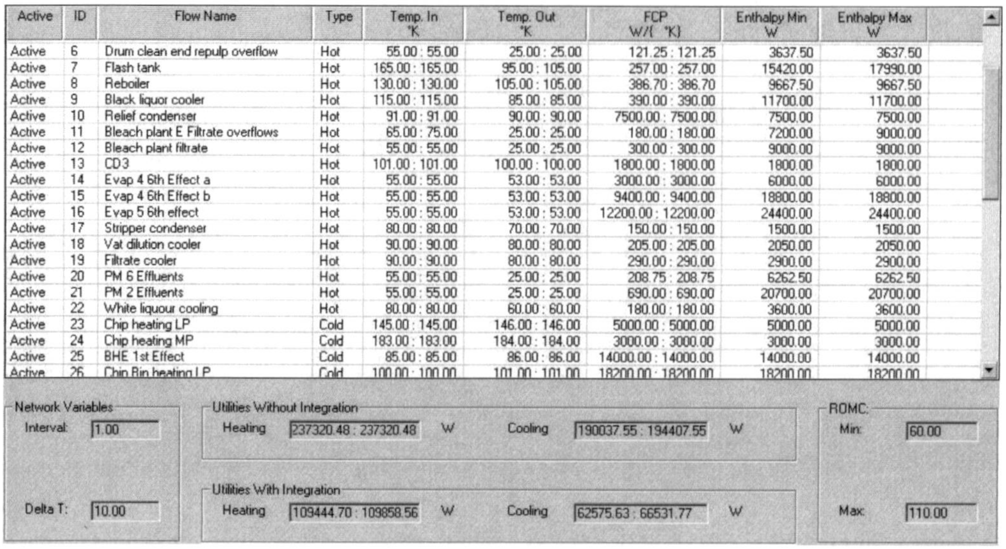

Active	ID	Flow Name	Type	Temp. In °K	Temp. Out °K	FCP W/(°K)	Enthalpy Min W	Enthalpy Max W
Active	6	Drum clean end repulp overflow	Hot	55.00 : 55.00	25.00 : 25.00	121.25 : 121.25	3637.50	3637.50
Active	7	Flash tank	Hot	165.00 : 165.00	95.00 : 105.00	257.00 : 257.00	15420.00	17990.00
Active	8	Reboiler	Hot	130.00 : 130.00	105.00 : 105.00	386.70 : 386.70	9667.50	9667.50
Active	9	Black liquor cooler	Hot	115.00 : 115.00	85.00 : 85.00	390.00 : 390.00	11700.00	11700.00
Active	10	Relief condenser	Hot	91.00 : 91.00	90.00 : 90.00	7500.00 : 7500.00	7500.00	7500.00
Active	11	Bleach plant E Filtrate overflows	Hot	65.00 : 75.00	25.00 : 25.00	180.00 : 180.00	7200.00	9000.00
Active	12	Bleach plant filtrate	Hot	55.00 : 55.00	25.00 : 25.00	300.00 : 300.00	9000.00	9000.00
Active	13	CD3	Hot	101.00 : 101.00	100.00 : 100.00	1800.00 : 1800.00	1800.00	1800.00
Active	14	Evap 4 6th Effect a	Hot	55.00 : 55.00	53.00 : 53.00	3000.00 : 3000.00	6000.00	6000.00
Active	15	Evap 4 6th Effect b	Hot	55.00 : 55.00	53.00 : 53.00	9400.00 : 9400.00	18800.00	18800.00
Active	16	Evap 5 6th effect	Hot	55.00 : 55.00	53.00 : 53.00	12200.00 : 12200.00	24400.00	24400.00
Active	17	Stripper condenser	Hot	80.00 : 80.00	70.00 : 70.00	150.00 : 150.00	1500.00	1500.00
Active	18	Vat dilution cooler	Hot	90.00 : 90.00	80.00 : 80.00	205.00 : 205.00	2050.00	2050.00
Active	19	Filtrate cooler	Hot	90.00 : 90.00	80.00 : 80.00	290.00 : 290.00	2900.00	2900.00
Active	20	PM 6 Effluents	Hot	55.00 : 55.00	25.00 : 25.00	208.75 : 208.75	6262.50	6262.50
Active	21	PM 2 Effluents	Hot	55.00 : 55.00	25.00 : 25.00	690.00 : 690.00	20700.00	20700.00
Active	22	White liquour cooling	Hot	80.00 : 80.00	60.00 : 60.00	180.00 : 180.00	3600.00	3600.00
Active	23	Chip heating LP	Cold	145.00 : 145.00	146.00 : 146.00	5000.00 : 5000.00	5000.00	5000.00
Active	24	Chip heating MP	Cold	183.00 : 183.00	184.00 : 184.00	3000.00 : 3000.00	3000.00	3000.00
Active	25	BHE 1st Effect	Cold	85.00 : 85.00	86.00 : 86.00	14000.00 : 14000.00	14000.00	14000.00
Active	26	Chip Bin heating LP	Cold	100.00 : 100.00	101.00 : 101.00	18200.00 : 18200.00	18200.00	18200.00

Network Variables
Interval: 1.00
Delta T: 10.00

Utilities Without Integration
Heating 237320.48 : 237320.48 W Cooling 190037.55 : 194407.55 W

Utilities With Integration
Heating 109444.70 : 109858.56 W Cooling 62575.63 : 66531.77 W

ROMC:
Min: 60.00
Max: 110.00

Fig. 5. Effect of h7 and h11 streams on the energy targets

Table 2. Set of streams to be controlled for minimum heating utility consumption

	Name	Initial		New	
Hot Steams					
		Supply Temp °C	Supply Temp °C	Supply Temp °C	Supply Temp °C
h2	Blow Accumulator Drain	80	20	85	25
h6	Drum Clean end Repulp over flow	60	20	55	25
h7	Flash Tank	160	100	165	95-105
h8	Reboiler	125	110	130	105
h9	Black liquor cooler	110	90	115	85
h11	Bleach Plant E Filtrate overflows	70	20	65-75	25
h12	Bleach Plant Filtrate	60	20	55	25
h20	PM 6 Effluents	60	20	55	25
h21	PM 2 Effluents	60	20	55	25
Cold Streams					
C22	Deaerator Feed water heating	15	145	10	140
C23	White Liquor heating	80	170	85	165
C24	Black Liquor heating	80	170	85	165

It is interesting to note here that we can sacrifice a little bit in realizing the global minimum heating utility to attain a "good enough" local minimum heating utility consumption and in the same time attain the global minimum cooling utility. For a minimum heating utility consumption target of 109.8 MW (which is only 0.35% higher than the global minimum) we can realize global minimum cooling utility target and still keep an important saving in the heating utilities consumption by about 8%.

This can be done through the control of the target temperature of h7 at 105°C (instead of 100°C) and the supply temperature of h11 at 65°C (instead of 70°C) as shown in Fig. 6 below.

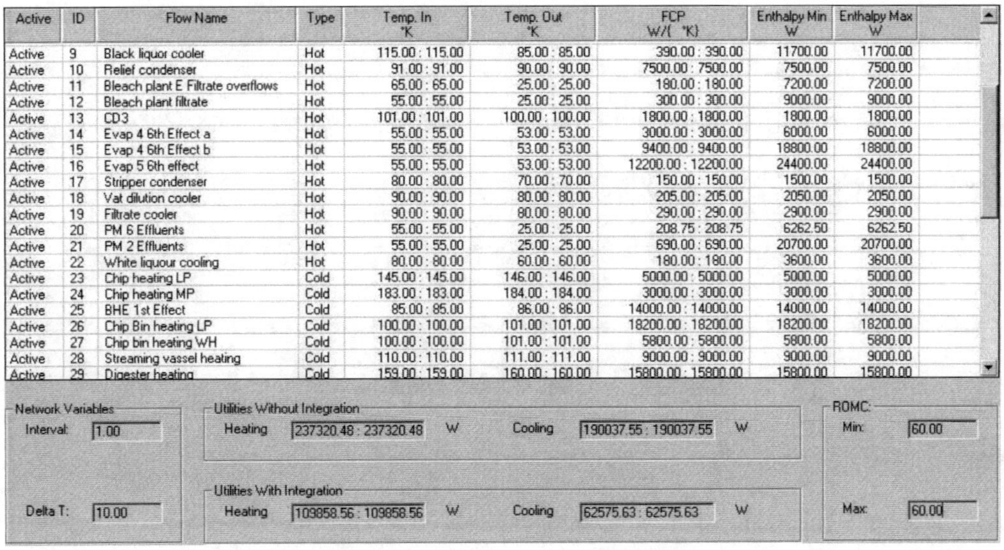

Active	ID	Flow Name	Type	Temp. In °K	Temp. Out °K	FCP W/(°K)	Enthalpy Min W	Enthalpy Max W
Active	9	Black liquor cooler	Hot	115.00 : 115.00	85.00 : 85.00	390.00 : 390.00	11700.00	11700.00
Active	10	Relief condenser	Hot	91.00 : 91.00	90.00 : 90.00	7500.00 : 7500.00	7500.00	7500.00
Active	11	Bleach plant E Filtrate overflows	Hot	65.00 : 65.00	25.00 : 25.00	180.00 : 180.00	7200.00	7200.00
Active	12	Bleach plant filtrate	Hot	55.00 : 55.00	25.00 : 25.00	300.00 : 300.00	9000.00	9000.00
Active	13	CD3	Hot	101.00 : 101.00	100.00 : 100.00	1800.00 : 1800.00	1800.00	1800.00
Active	14	Evap 4 6th Effect a	Hot	55.00 : 55.00	53.00 : 53.00	3000.00 : 3000.00	6000.00	6000.00
Active	15	Evap 4 6th Effect b	Hot	55.00 : 55.00	53.00 : 53.00	9400.00 : 9400.00	18800.00	18800.00
Active	16	Evap 5 6th effect	Hot	55.00 : 55.00	53.00 : 53.00	12200.00 : 12200.00	24400.00	24400.00
Active	17	Stripper condenser	Hot	80.00 : 80.00	70.00 : 70.00	150.00 : 150.00	1500.00	1500.00
Active	18	Vat dilution cooler	Hot	90.00 : 90.00	80.00 : 80.00	205.00 : 205.00	2050.00	2050.00
Active	19	Filtrate cooler	Hot	90.00 : 90.00	80.00 : 80.00	290.00 : 290.00	2900.00	2900.00
Active	20	PM 6 Effluents	Hot	55.00 : 55.00	25.00 : 25.00	208.75 : 208.75	6262.50	6262.50
Active	21	PM 2 Effluents	Hot	55.00 : 55.00	25.00 : 25.00	690.00 : 690.00	20700.00	20700.00
Active	22	White liquour cooling	Hot	80.00 : 80.00	60.00 : 60.00	180.00 : 180.00	3600.00	3600.00
Active	23	Chip heating LP	Cold	145.00 : 145.00	146.00 : 146.00	5000.00 : 5000.00	5000.00	5000.00
Active	24	Chip heating MP	Cold	183.00 : 183.00	184.00 : 184.00	3000.00 : 3000.00	3000.00	3000.00
Active	25	BHE 1st Effect	Cold	85.00 : 85.00	86.00 : 86.00	14000.00 : 14000.00	14000.00	14000.00
Active	26	Chip Bin heating LP	Cold	100.00 : 100.00	101.00 : 101.00	18200.00 : 18200.00	18200.00	18200.00
Active	27	Chip bin heating WH	Cold	100.00 : 100.00	101.00 : 101.00	5800.00 : 5800.00	5800.00	5800.00
Active	28	Streaming vassel heating	Cold	110.00 : 110.00	111.00 : 111.00	9000.00 : 9000.00	9000.00	9000.00
Active	29	Digester heating	Cold	159.00 : 159.00	160.00 : 160.00	15800.00 : 15800.00	15800.00	15800.00

Fig. 6. Optimal plant operating conditions for minimum energy consumption

6. Conclusion

Real-time heat integration in pulp & paper industry renders new opportunity for waste energy recovery and can be economically justified. An average saving of 8% in hot utility consumption can be realized through the proper control of a specified set of plant operating conditions. In a typical pulp &paper plant instantaneous savings in the hot utility consumption under minor process disturbances can reach about 19%. While real-time heat integration is promising from energy cost saving point of view it needs to be considered in the context of the total cost impact on the plant's energy system. **TEM_icons™** software can be used to systematically target for both best and worst case scenarios for hot and cold utilities consumptions under all possible combinations of process disturbances. It can also used to define without enumeration the set of process conditions that need to be closely controlled to keep minimizing plant's utility bill of highly disturbed processes.

Reference

[1] MBendi Business website, Africa's leading business website, World: Pulp & Paper industry, www.mbendi.co.za/indy/pulp/p0005.htm

[2] Forest Conservation Portal, Forest Conservation Archives, http://forests.org/articles/reader.asp?linkid=36525

[3] IEA Technology agreements, Pulp & Paper, www.iea.org/dbtw-wpd/Textbase/techno/iaresults.asp?la=Pulp%20and%20Paper

[4] M.B. Noureldin, Improved system and computer software for modeling energy consumption, New Zealand Patent No.527244, 2003

[5] M.B. Noureldin, A.K. Hasan, Interval constraint satisfaction-based software for energy optimization in chemical plants, Proceedings of the 8th International conference on Production Engineering Design and Control, Alexandria, Egypt, 27-29 December 2004.

[6] M.B. Noureldin, E.B. El-Halwagi, Interval-based targeting for pollution prevention via mass integration, Computers and Chemical Engineering 23 (1999) 1527-1543

[7] M.B. Noureldin, J.E. Swan, Computer-Aided design software for energy optimization through interval constraint logic propagation, Proceeding of MDP-8, Cairo University Conference on Mechanical Design and Production, Cairo, Egypt, January 2004.

[8] M.B. Noureldin, "TEM_icons™ 1.2 User's manual" Report, Department of Materials and Process Engineering, University of Waikato, Hamilton, New Zealand, 2003.

[9] M.B. Noureldin, A.K.Hasan, Global energy targets and optimal operating conditions for waste energy recovery in Bisphenol-A plant, Applied Thermal Engineering 26 (2006) 374-381

95d Technology Assessment of Rejecting Process Heat

C.B. Panchal, Energy Systems Division, Argonne National Laboratory, 9700 S Cass Ave, Argonne, IL 60439

The value of an efficient heat rejection system is generally not considered in the original design. Many petrochemical plants and refineries are operating on a tight cooling water supply, which could impact the future expansion of capacity. An assessment of heat rejection in a hypothetical refinery is evaluated and various technologies for rejecting and/or utilizing the process heat are presented. They include waste heat utilization by upgrading, bottoming cycle to produce power, and heat rejection with air cooling. The status of underlying technologies for new generation of heat rejection systems is discussed, and major technical as well as economic issues are identified.

95e - Spiral Flow Heat Exchanger for Heat Recovery from Gases

Pramod Madhukar Deshpande

Full paper on next page

Introduction

Energy recovery is the prime requirement today to optimize energy consumption. The maximum utilization of thermal energy is achieved by properly designed Heat Exchangers, and selection of temperature program. The selection of heat exchanger geometry is also important for particular heating /cooling application. There are various heat exchanger geometries exist and to be selected based on applications. However, the Horizontal spiral heat exchanger is a unique geometry of compact heat exchange for gas/liquid; vapor/liquid heat transfer applications.

The different designs of Shell and tube heat exchanger exist for heat recovery from Diesel Generating Power plants of capacity 1000 **KVA** and above. However, for medium to small size Diesel generating power plants, the shell and tube heat exchanger is ineffective. The flue gas quantity generated by medium to small Diesel generating power plants is low. The Shell and tube heat exchanger geometry is ineffective due to low flue gas velocity in the heat exchanger. To avoid the effect of suit formation it is recommended to use tubes of bigger size. In case, of medium to small size Diesel generating power plants. The horizontal spiral tube heat exchanger geometry provides the solution for such application. The gas velocity, velocity of liquid and its boundary layer, predominantly governs the heat recovery in flue gas/ hot gases by liquid. Also, the heat recovery is directly depending on the surface area density; it is meter square of heat transfer surface area per meter cube. The spiral tube heat exchanger provides the solution for such applications.

Horizontal spiral tube heat exchanger

The horizontal spiral tube heat exchanger is a geometry in which the tube of desired size is made in to the spiral as shown in Fig 1. The predefined distance is maintained between the two successive turns of spirals. There could be number of turns, however it is restricted by the flow patterned developed in spiral curved tube. The maximum no of turn depends on the property of fluid such as density, viscosity as it governs the turbulence.

Fig 1. Horizontal spiral tube heat exchanger Coil

Flow through curved tube

The flow of fluid through a curved coil /pipe creates a secondary circulation of fluid called double –eddy or Dean Effect and it is taking place in plane perpendicular to main flow. When fluid passes through 90^0 bends a secondary flow is established. The development of secondary flow is independent of the laminar or turbulent flow. This rotating motion of liquids flowing through coiled tubes develops a characteristics swirling action. The frictional resistance of the tube walls and action of centrifugal force combine to produce this rotation. The degree of secondary flow produced is a function of ratio of curvature of coil to the tube diameter(r/D)and flow velocity. The faster the velocity of fluid for a given (r/D) ratio the greater the centrifugal force and more predominant is the secondary flow.

Pressure drop

The pressure drop thorough the curve tube is higher than straight tube. If the curved tube/coil of certain spiral diameter is made then the pressure drop in the coil increases linearly with number of turns of coil. It is expressed as pressure drop in one turn of coil multiplied by number of turns of coil. However, in case of horizontal spiral coil as the diameter of spiral increases the pressure drop in that particular turn decreases. This result in lower pressure drop as compared to vertical spiral coil at the same time ensures the secondary flow in coil. The nature of secondary flow will depend on density and viscosity of fluid. The flow pattern in curved tube as shown in Fig 2

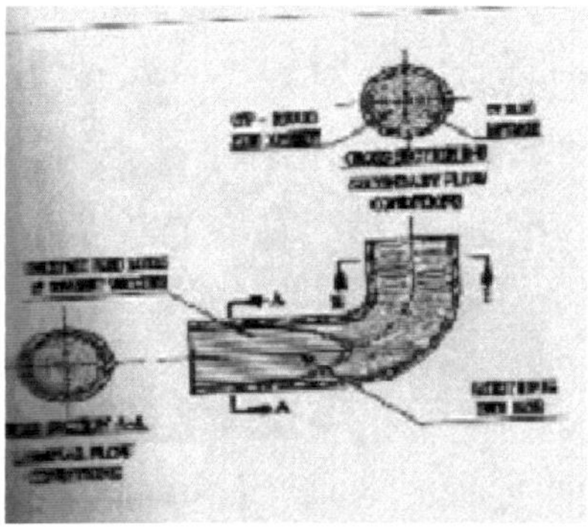

Fig 2 Flow through curved tube of 90 deg bend
(Secondary Flow - Dean Effect)

Heat Transfer

The secondary flow of fluid through curved tubes and in particular horizontal spiral tube heat exchanger ensure the higher rate of heat flow through the fluid. The mechanical agitating is induced due to fluid motion hence higher heat transfer rate achieved without any external mechanical induced devise in tube.

Waste heat from D.G.Set

Due to the increased demand of power the medium and small diesel engine power plant are installed extensively for their energy requirement. Only the part of total fuel energy supplied to the internal combustion engine is converted into useful work. The work is delivered at the crankshaft and rest of the fuel energy is rejected as

- Heat rejected through Exhaust Gases
- Heat rejected through coolant
- Heat from engine boundaries due to radiation, convection and to small extend the conduction.

The percent of fuel energy to power is governed by physical and design characteristics of diesel engine, type of fuel, its physical characteristics, and cooling system. The heat loss from the flue gases is predominant; In general about 40% of energy input is converted into useful work as power. 35 % is lost as exhaust heat 20 % to coolant and 5 % as convective and radiation heat loss. It is seen that major loss of energy is through the exhaust gases and cooling water. To improve the energy efficiency it is recommended to recover the heat either from exhaust or from cooling water.

Flue gas temperature

The exhaust flue gas temperature of diesel engine power plant for small to medium capacity power plant varies from 380 to 350 ^{0} C depending power output. The flue gas temperate is much higher and it is possible to reduce the flue gas temperature and recover the heat.

Heat Recovery

The recovery of heat from flue gases is to be done by a suitably design heat exchanger; the horizontal spiral tube heat exchanger is more suitable for this application due the compact design. Compact ness factor; meter square of surface area per meter cube of heat exchanger volume is higher as compared to shell and tube heat exchanger. The heat can be recovered by water by rising its temperature and hot water can be used for varied process application.

Efficiency

The fuel energy efficiency is improved and the heat losses from flue gases are reduced from 35 % to 10-15 %. That is 20 % additional utilization of fuel energy.

Heat Exchanger

The heat exchanger is the process heat transfer equipment. The type of heat exchanger governs the effectiveness of heating and cooling. The different geometries of heat exchanger are possible however the proper selection required based on application. The flue gas heat recovery is a heat transfer problem of gas / liquid heat transfer. The

Horizontal Spiral tube heat exchanger

The horizontal spiral tube heat exchanger is a geometry in which the tube coil is made in one plane. The coil is developed in such a way that the spiral coil diameter is continuously increasing as shown in figure 3.

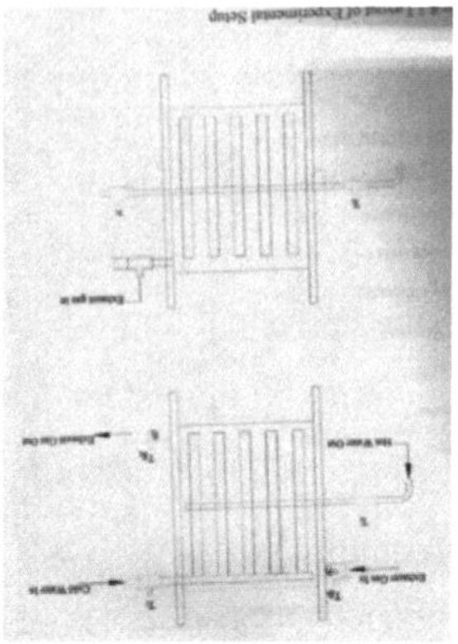

Figure 3 Heat Exchanger Geometry

The spiral shape of tube results in continues change in direction of fluid flow results in local secondary flow. The secondary flow depends on the radius of spiral and fluid properties. The development of secondary flow in tube brakes the boundary layer and it result in

- High share rate at wall
- Increased turbulence
- Decreased Boundary layer

In the horizontal spiral tube heat exchanger, the outside flow pattern also gets modified. The heat exchanger is as cross flow heat exchanger. The heat transfer rate an flux decides the effectiveness of heat exchanger. The cross flow heat exchanger ensures higher effectiveness in services for gas / liquid heat transfer as in case of air-cooled systems or refrigeration systems. The horizontal spiral tube heat exchanger is a modular design. The heat transfer area can be increased either by increasing the number of turns of spiral or adding the number of spiral coils. The correlations for estimating pressure drop are reported in literature; the effect of change of direction of fluid flow in tube and its effect on heat transfer are been studied in very few cases. Hardly any data is available to analyze enhancement of heat transfer but the geometries has wide applications due to its inherent advantages.

Design of HSTHE

The horizontal spiral tube heat exchanger heat transfer surface area is estimated base on analysis of double pipe and vertical spiral tube heat exchanger design analysis. The earlier studies conducted by us on horizontal spiral tube heat exchanger geometry indicates that there is an enhancement of overall heat transfer coefficient. The HSTHE is investigated as under

Material of Construction

The spiral coil of horizontal spiral tube heat exchanger can be made of varied material; the ductile material like copper is ideal for preparing coil, however the other material like stainless steel are possibly to make as spiral coil with pretreatment. The shell of heat exchanger can be of stainless steel or mild steel depending on the duty conditions. For this investigation the stainless steel tube is coil with mild steel is used for investigation.

Experimental Study

The heat recovery from the diesel power generating set has been studied experimentally on the diesel power generator. The diesel power generating set is RU3TR with a speed variation of 1000 to 2000 rpm. The D.G .power plant work on the high speed diesel. The heat exchanger consist of 5 coils of 8 mm tube of stainless steel 316 the other specifications are as under

HSTHE specification

1. Length of tube - 1000 mm
2. Number of coils - 5 no.
3. Diameter of coil - 140 mm
4. Diameter of shell - 160 mm ID
5. Shell Thickness - 5 mm
6. Heat transfer surface area - 52 mm^2
7. Water inlet/outlet Header - 250 mm
8. Design Temperature
    - Max flue gas - 400 ^{0} C
    - Water inlet - 27 ^{0} C

The ambient temperature water (27 ^{0} C to 29 ^{0} C) is allowed to pass through the spiral coil through the inlet header. The hot water is collected from the outlet header. The flue gas is allowed to pass over the bank of spiral coil tube bundle results in cross flow heat exchanger. The thermocouples are installed on inlet and outlet of hot as well as cold fluid to measure the temperature. The pressure drop on waterside and flue gas side is measured with the help of manometers. The cold-water flow is measured continuously by pre calibrated volumetric flask. The flue gas quantity is estimated based on specific fuel consumption of D.G. power plant and difference in tank level of oil tank. The heat recovery studies are carried at different fuel consumption so that the flue gas quantity and flue gas outlet temperature from D.G. power plant is varied. The heat exchanger is installed at out let of the Diesel generating power plant. The temperature on flue gas and waterside is measured with thermocouples. The water is allowed to pass through coil.

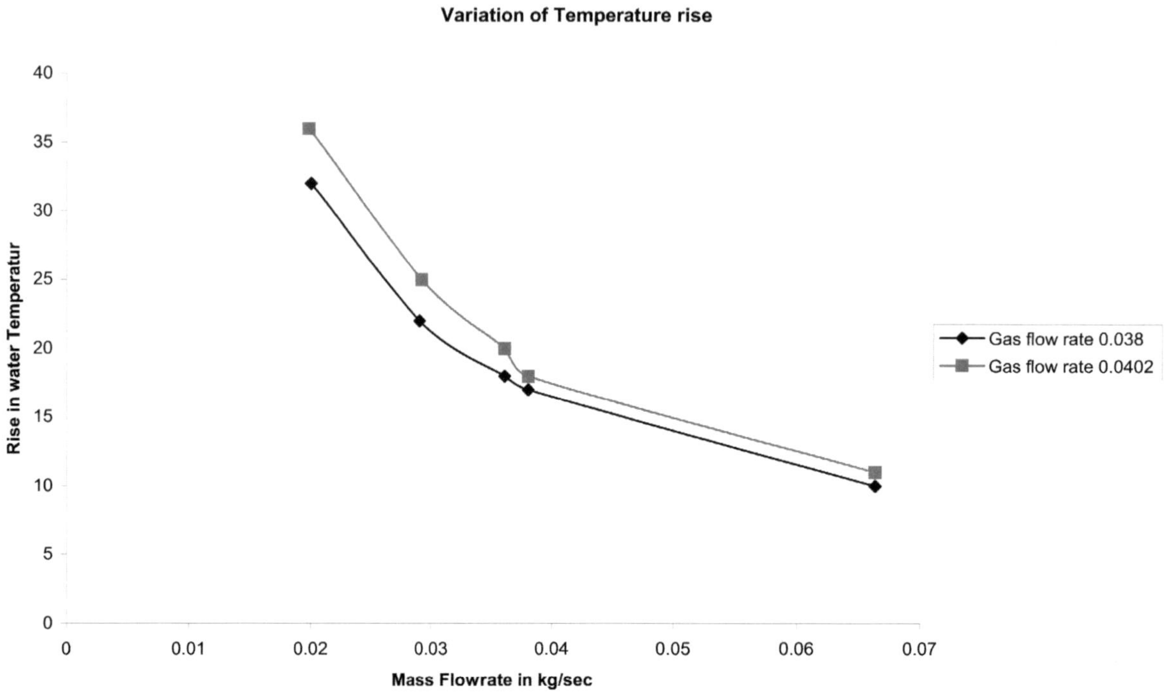

Figure 4 a) Performance of Heat recovery system

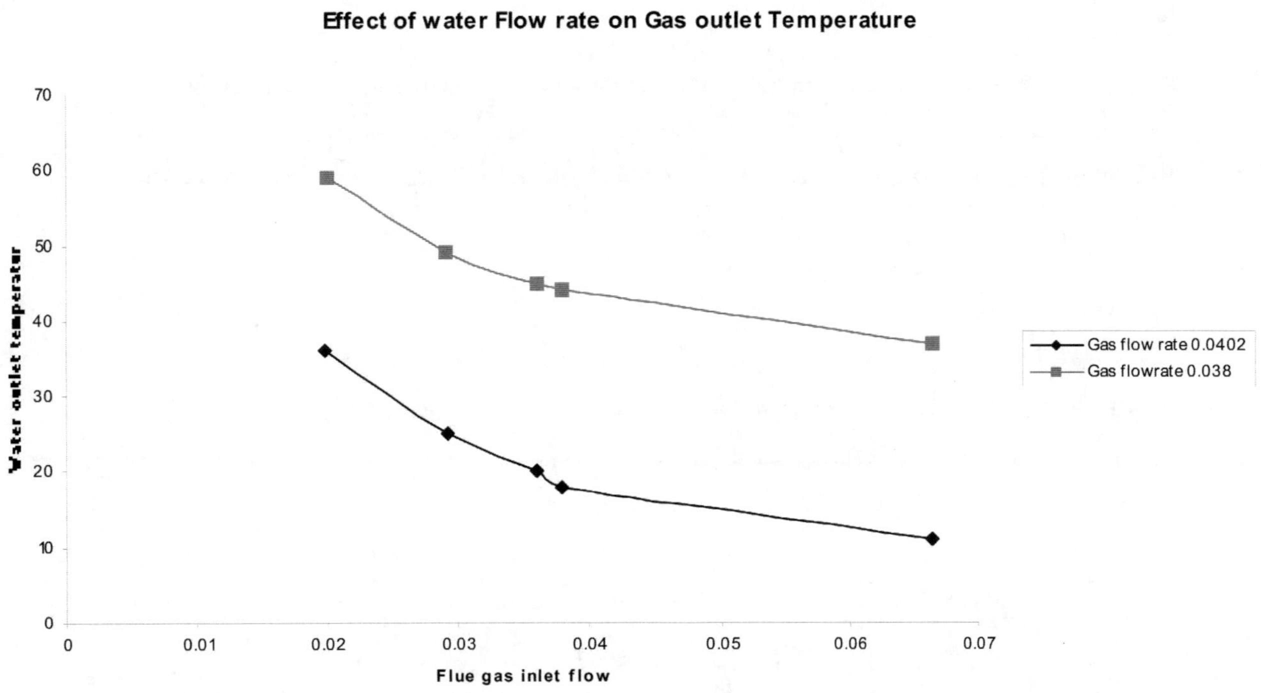

Figure 4b Variation of Outlet Temperature

Result analysis

The hydrodynamic study heat exchanger suggest that the pressure drop varies with the water velocity; however the pressure drop is maximum at a velocity indicating the maximum turbulence exists at certain velocity. The detail investigation with varied number of coil turns is under investigation.

The study of results of heat recovery from heat exchanger suggest that with variation of water flow rate the rise in temperature of water get varied. The quantity of heat transfer remains constant.

The figure 4a and Figure 4b explains the variation of water outlet temperature and temperature rise for various cooling water flow rates. The studies are carried out for different fuel consumption.

Advantages

The major advantages are fast response to fluctuating load due to compactness. The design and construction eliminates the support structure in case of condensers for vertical tall vessels like distillation columns, reactors. Lower space requirement; modular construction helps in modifying the heating cooling duty.

HSTHE applications

The horizontal spiral tube heat exchanger accommodates more heat transfer are per cubic meter which helps in modulating the heat load. The compactness construction help in

- Localized heating/ cooling in reactor
- Enhancing the heat load in case of condensation duties
- Condensation duties of small to medium size distillation systems
- Cooling the effluent gas by suspending the coils in flue gas path
- Cooling the air by chilled water.

The above are some of the applications.

Conclusions

The horizontal spiral tube heat exchanger has higher pressure drop as compared to double pipe heat exchanger, however there is an enhancement in heat transfer rate, which off sets the disadvantage of pressure. The compact construction is suitable for heat recovery from gases. .

96a Methanol Powered Microchannel Fuel Processing Systems

Robert Dagle[1], Daniel Palo[2], and Jamelyn D. Holladay[1]. (1) Pacific Northwest Laboratory, 902 Battelle Blvd, PO Box 999, Richland, WA 99352, (2) Microproducts Breakthrough Institute, Pacific Northwest Laboratory, PO Box 2330, Corvallis, OR 97339

Battelle, Pacific Northwest Division (Battelle) has been active in developing fuel processors and power systems at various power levels to meet the growing needs of the U.S. military. For low power applications, methanol fuel processors have been developed at outputs ranging from sub-watt to several hundred watts. For higher power, fuel processor systems capable of operating on a wide variety of fuels, including synthetic JP8, have been developed. Methanol fueled small power systems (<1 kw) will be discussed in this presentation. Each of these systems operates employs Battelle's proprietary methanol steam reforming catalyst, CO cleanup catalyst, and patented microchannel architecture. Microchannels provide rapid heat and mass transfer in the process streams, while proprietary catalysts offer high conversion activity and excellent product selectivity. These technologies enable small, light weight devices that provide a reliable stream of hydrogen-rich gas to power small proton exchange membrane (PEM) fuel cells for various military applications.

96b Comparison of Adiabatic Gas-Liquid Two-Phase Flows in Microchannels and Minichannels

Masahiro Kawaji, Department of Chemical Engineering and Applied Chemistry, University of Toronto, 200 College Street, Toronto, ON M5S 3E5, Canada and Hideo Ide, Department of Mechanical Engineering, Kagoshima University, Kagoshima, Japan.

Gas-liquid two-phase flows in small channels occur in micro heat exchangers, PEM fuel cells, micro reactors and other energy-related equipment. To better understand the flow characteristics and design more efficient devices and equipment, the adiabatic gas-liquid two-phase flow characteristics in small diameter circular and square channels have been analysed by examining the nitrogen gas-water two-phase flow data. Recent two-phase flow experiments conducted using microchannels with a diameter of 20 ~ 500 microns have shown both significant differences and some similarities with those measured in minichannels with a diameter of 500 microns ~ 5 mm. The microchannel flows have also been shown to be sensitive to the inlet geometry and the method of gas-liquid injection. In this work, the two-phase flow patterns, void fraction and two-phase pressure drop data obtained in small channels of 100 microns ~ 5 mm diameter have been analysed and compared to further clarify the effect of channel diameter on various two-phase flow parameters, including a wave velocity. In circular channels, a ring flow pattern has recently been considered to be a characteristic of microchannel two-phase flow, however, it could also be observed in minichannels with a diameter greater than 1.0 mm. This flow pattern is a type of an annular flow with a wavy interface profile with the wave peaks spaced apart at a regular interval. The wave velocity data obtained in an annular flow pattern showed that it increases steadily with a decreasing diameter from 5 mm to about 1.5 mm, and then starts to decrease as the channel diameter is further reduced. The radius of curvature of the channel affects the wave velocity as the significance of surface tension force dramatically rises as the channel diameter is reduced.

The void fraction data obtained in microchannels and minichannels showed significant differences depending on the channel cross section and inlet geometry. For the microchannel with a diameter of 100 microns, the effect of the inlet geometry on the void fraction was seen to be quite strong, while the minichannels showed a much smaller effect of inlet geometry on the void fraction. In square channels, a ring flow pattern observed in microchannels could also be observed in a 1.0 mm x 1.0 mm channel, however, the conditions under which this flow pattern could occur was limited in comparison with the circular channel. The liquid tended to flow along the four corners, reducing the liquid film thickness in both microchannels and minichannels. As a result, the void fraction data in square channels showed lower values in comparison with those obtained in circular channels. The two-phase pressure drop data also showed an increasingly stronger effect of the inlet geometry as the channel diameter was reduced.

Modeling and evaluation of an integrated microfluidic fuel processor for miniature power sources

Dongying Qian, Ronald S. Besser, Adeniyi Lawal
Stevens Institute of Technology, Hoboken, NJ, USA

Abstract

New lighter-weight, longer-lasting portable power sources are in demand to meet the special requirements for laptop computer, cellular phone, military and homeland security use. A microreactor technology based fuel cell can achieve high power and energy density by directly transforming chemical energy into electrical energy, thus providing an alternative to current secondary batteries. Proton exchange membrane fuel cell (PEMFC) is an attractive technology for low power level applications. In PEMFC applications, an on-board fuel processor would be desirable due to the difficulty associated with the storage and refilling of hydrogen. However, there are a number of technical challenges that need to be addressed for the development of portable fuel processor to become commercially viable. Miniaturization in terms of size and weight, fuel processor lifetime, and material and energy management are some of the most critical issues. In this presentation, we describe the design and simulation of a micro-reactor-based fuel processor on silicon wafers for a 20 W PEMFC with steam reforming of methanol. A hardware integration scheme, which incorporates four major components: vaporizer, reformer, preferential oxidation reactor (PrOx) and combustor, several internal heat exchangers, and some internal thermal barriers was proposed. A complete two-dimensional CFD simulation of the integrated system involving flow, heat and mass transfer as well as chemical reactions was performed. Detailed information such as species concentration profiles, reaction extent, temperature profile and energy usage was obtained. The size of each major equipment component including the internal heat exchangers was determined, as well as the thickness of the insulators separating them.

Keywords: Microreactor; Integration; Fuel Cell; Modeling

Introduction

For decades, fuel cells have attracted increasing attention as potential energy sources due to high energy efficiency and significant emission reductions [1-3]. Most of the world's leading automotive manufacturers have fuel cell vehicle research and development programs. Nowadays, this interest has been extended to the smaller scale fuel cells. The advances of portable electronic devices for residential use such as laptop computer and cellular phone, and also for military and intelligence use such as advanced sensing, imaging, data analysis, communications and weaponry capabilities require high-density light-weight power sources. The application of microreactor technology in the field of fuel cell processing provides the opportunities for new portable battery sources. Microreactor technology can facilitate highly compact power sources through integration of several unit operations with microstructured sensors, actuators and other "balance of plant" (BOP) devices to form power-plant-on-chip. Among several lower power range portable fuel cell applications, the proton exchange membrane (PEM) fuel cell is viewed as a commercially viable technology for those applications of low power levels. PEMFC technology provides high power density while utilizing relatively low loading of precious metal electrocatalysts within the fuel cell; it is simple in design and operated in mild conditions while providing instant power [4-7]. However, since it is difficult to store and refill hydrogen, many researchers are developing PEM fuel cell technology with the installation of a hydrogen generation plant on board. The choice of fuel for the fuel cell stacks is also a key issue regarding the commercialization of the technology. Methanol steam reforming fuel cell technology is recognized as suitable for those applications that require 20 watts and above due to the high energy density of methanol (~5.6 kWh/kg compared to lithium-ion batteries ~0.12 kWh/kg) [8].

The methanol steam reforming reaction is a highly developed and thoroughly studied process and it is possible to yield a product gas containing up to 75% hydrogen while maintaining a high selectivity towards carbon dioxide [7]. Compared to other sources of hydrocarbon fuels such as diesel, methanol steam reforming fuel processing will significantly reduce the amount of CO in the production stream (~1-2%), which can simplify the carbon monoxide cleanup step by eliminating the water gas shift (WGS) reaction component. Because the PEMFC is operating at low temperature (60~80 ^{o}C), the electrocatalyst at the anode is extremely sensitive to CO: a small amount of CO will poison and deactivate the catalyst rapidly. The concentration of CO in the H_2 stream to the FC anode should be less than 10 ppm. A preferential oxidation (PrOx) reaction component, which selectively oxidizes carbon monoxide into carbon dioxide, is generally required to serve this purpose. On another hand, the feed of methanol and water is in liquid state, and needs to be evaporated before entering the steam reforming reactor, so an evaporating device is required. The steam reforming reaction is endothermic and as result significant amount of heat has to be provided to maintain the reaction. A catalyst combustor is generally used in order to supply the necessary heat for the reforming reaction. The combustor will burn the fuel cell anode off gas (AOG) for combustion and be operated at relatively mild temperature with the catalyst for reducing the emission. Therefore, a methanol steam reforming fuel processor will include four main components, Vaporizer, Reformer, PrOx, Combustor, and the corresponding heat exchangers, sensors, actuators and other BOP devices.

For the portable applications, miniaturization in terms of size and weight, long-lasting lifetime and the energy management are the critical issues. In fact, these are all the challenges and major hurdles in the development of portable fuel cell processes. In order to achieve these goals, an appropriate integration of components, especially with the fuel processing units is demanded. Many researchers have proposed different hardware integration schemes in PEM fuel cell studies [9-12]. However, the majority of them focus on the study of bigger-scale power range, and integration is typically on two processing units. The integration of PEM fuel cell at low power range is still in its infancy. In the portable power range, integration of a variety of components is required, including fuel processing units, heat exchangers, sensors and actuators, etc. It also requires the assembly of different kinds of materials. With the development of microfabrication technology, integration of a variety of components and different materials in a small device can be realized.

An integration scheme for a 20W PEM fuel processor involving the steam reforming of methanol fuel is proposed and examined here. This power range will sustain the energy requirement for soldier platform while achieving the goal of high energy density, light weight and compact volume. The integration system incorporates four major FC process components: Vaporizer, Reformer, PrOx and Combustor, several internal heat exchangers, and internal thermal barriers for the maintenance of different operating temperatures of the adjacent processing units. A complete two-dimensional CFD simulation on the integrated system considering flow, heat and species mass transfer as well as chemical reactions was performed.

Detailed information such as species concentration profiles, reaction extent in each reactor unit and temperature profiles inside the integrated system was obtained. The size of each main component, internal heat exchangers, and thermal barriers was predicted.

Process Simulation

Our first step for the 20W PEM fuel processor design is to simulate and optimize the FC system by considering a plant model in which the main components are modeled with some simplified but rational assumptions. The objectives of this simulation are to obtain processing material and energy balance, optimize the operating conditions of each subsystem as well as the overall system, and analyze the possible integration of different components with respect to their heat demand or heat production. A process simulator Aspen Plus v12.1 is applied for this task.

We consider here 4 key fuel processing components, "vaporizer (VAPZ)", "steam reformer (SR)", "preferential oxidation reactor (PrOx)", "combustor (COMB)", 1 fuel cell stack (FC), and several heat exchangers. The fuel processor will convert methanol into H_2 rich stream with <10 ppm CO. The entire stream feed requirements (flow rate, temperature, pressure, compositions), operating conditions of all process equipment components (temperature, pressure), and energy management strategy (thermal process and hardware integration) need to be carefully selected and optimized. Based on literature data and intrinsic kinetic behavior of each reaction, an optimum process analysis and design of 20W fuel processor system has been realized. Figure 1 shows the overall material & energy utilization of this system from Aspen Plus simulation. In this system, hydrogen utilization in the fuel cell (FC) unit is assumed to be 70%, and the overall energy efficiency is 31%.

In the process simulation, we consider each reactor being operated under isothermal condition and at atmospheric pressure. Heat losses from all the components are not considered. The extra 4W output power from the fuel cell stack is considered to be consumed by FC processing internal use for the pumps and compressors. The unconsumed hydrogen from anode off-gas goes into the combustor for combustion which provides the heat for the entire system. From the material and energy management assessment, globally this system consumes methanol, water and air, produces water and carbon dioxide, and generates power and heat.

The global stream feed requirements, operating conditions of each process component and energy demands are clearly indicated in Figure 1. Several internal heat exchangers are considered for heat management. The main operating parameters of the system are briefly described below for the main components.

1. Fuel cell stack

The PEM fuel cell stack generates electrical power through electrochemical reaction of hydrogen and oxygen. The anode reactant gas is H_2 rich reformate stream which is provided by the fuel processor, and the cathode reactant gas is humidified air. Both electrodes are coated with porous carbon-supported platinum catalyst with a solid, electronically insulated polymer membrane as an electrolyte. The reaction in the anode catalyst layer is hydrogen oxidation, and the reaction in the cathode catalyst layer is oxygen reduction. The electron passes from the anode to the cathode through the electrode backing layer and the external load. The proton produced at the anode is transferred through the polymer electrolyte and participates in cathode reduction. The overall PEM fuel cell reaction is as follows:

$$H_2 + \frac{1}{2} O_2 \rightarrow H_2O + \text{electrical energy} + \text{waste heat} \quad (1)$$

The PEM fuel cell converts chemical energy directly into electrical energy, while producing heat and liquid water as byproducts. The waste heat produced by the exothermic reaction of hydrogen and oxygen has to be removed through some kind of cooling loop. However, in the fuel cell stack, not all the electron produced by the anode reaction can be consumed by the external load; the optimum stack efficiency is about 60%. The hydrogen supplied to the fuel cell stack cannot be fully utilized either. In our design, we assume 70% hydrogen utilization on the anode side and less than 50% O_2 utilization on the cathode side in order to prevent electrode polarization. The fuel cell stack is operating at 70 °C and 1 atm with 31% overall energy efficiency in the entire system. This fuel cell unit will produce 24 W power energy as well as 20 W waste heat.

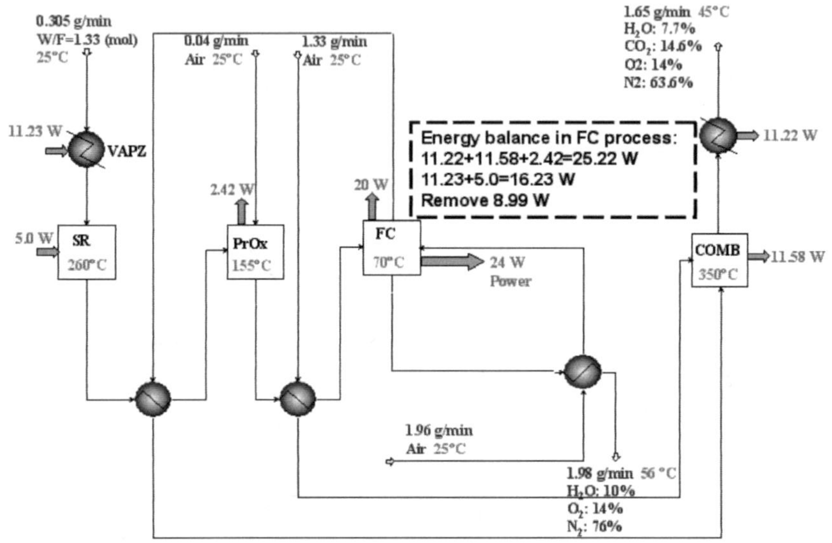

Figure 1 Material & energy balances in 20W fuel processor

2. Steam reformer

CuO/ZnO-based catalyst is generally used for catalytic steam-reforming of methanol. This process produces primarily hydrogen and carbon dioxide, but generates a small amount of carbon monoxide (~1-2%) as well. The main reactions taking place on this type of catalyst are as follows:

$$CH_3OH + H_2O = CO_2 + 3H_2 \qquad (\Delta H= +57 kJ/mol) \quad (2)$$

$$CH_3OH = CO + 2H_2 \qquad (\Delta H= +97 kJ/mol) \quad (3)$$

$$CO+H_2O=CO_2+H_2 \qquad (\Delta H= -41 kJ/mol) \quad (4)$$

where Eq. (2) is the desired steam reforming reaction, Eq. (3) is the methanol decomposition reaction, and Eq. (4) is the water-gas shift reaction. The methanol reformer is operated at 1-2 atm and 220-260 °C with excess amount of water in order to achieve high methanol conversion and low amount of CO [13]. The reformer is typically running at near thermodynamic equilibrium state. The reforming reaction Eq. (2) is relatively slow and it is an endothermic reaction, so the size of the reformer is large compared to the other units and a significant amount of heat needs to be provided for vaporizing the liquid fuel/water and compensating for the heat of the reaction. In our design, the reformer operates at 260 °C and 1 atm. The methanol conversion is close to 100% and CO concentration is 1.1% in outlet gas with water to methanol mole ratio of 1.33. The total amount of heat required for the vaporizer and reformer is 16.23 W.

3. Preferential oxidation of CO

As earlier stated, the main purpose of the PrOx reactor is to selectively oxidize CO to reduce its amount to <10 ppm. Pt/Al_2O_3 catalyst is commonly used and has been extensively studied in the literature because of its high CO conversion and stability at moderate temperatures (~200 °C). On this type of catalyst, the following main reactions are taking place simultaneously:

$$CO + \frac{1}{2}O_2 = CO_2 \qquad (\Delta H = -283 \text{ kJ/mol}) \qquad (5)$$

$$H_2 + \frac{1}{2}O_2 = H_2O \qquad (\Delta H = -242 \text{ kJ/mol}) \qquad (6)$$

$$CO_2 + H_2 = CO + H_2O \qquad (\Delta H = +41 \text{ kJ/mol}) \qquad (7)$$

where Eq. (5) is the desired CO preferential oxidation reaction, Eq. (6) is the hydrogen oxidation reaction, and Eq. (7) is the reverse water-gas shift reaction. Our center has extensively studied PrOx microreactor with thin-film Pt/Al_2O_3 catalyst [14-15]. Based on our experimental results, optimal operating conditions of 1 atm and 155 °C for PrOx are proposed. We can achieve <10 ppm concentration of CO in PrOx reactor. However, at high conversion of CO, any excess amount of O_2 fed in the stream for PrOx reaction would react with H_2 to form H_2O. The O_2 utilization for PrOx reaction is generally less than 50%. At this design condition, 2.42 W heat needs to be removed because of the exothermic reactions of Eqs.(5) & (6).

4. Combustor

Catalytic combustion in which the fuel/air mixture is converted heterogeneously over a catalyst is widely used in fuel cell applications because this process is operating at a mild condition and consequently the NOx emission can be significantly reduced. Since not all the hydrogen is consumed by the fuel cell stack, there is unreacted hydrogen available in a mixture with mainly water and carbon dioxide in the anode off gas (AOG). This AOG can be burned in a combustor to supply the heat. The reaction taking place in the combustor is the following:

$$H_2 + \frac{1}{2}O_2 = H_2O \qquad (\Delta H = -242 \text{ kJ/mol}) \qquad (8)$$

Pt/Al_2O_3 catalyst is assumed to be used here. The reaction temperature is 350 °C and the pressure is 1 atm. An excess amount of O_2 is supplied in the air stream, and we assume that this process operates under thermodynamic equilibrium condition. When the exhaust gas is cooled down to 45 °C, the total amount of heat available is 11.58+11.22=22.8 W. This amount of heat is enough to supply the heat requirements of the vaporizer and reformer. Therefore, there is no need to provide additional fuel into the combustor.

Hardware integration scheme

From the process analysis, we know that the above fuel processor/cell system is able to generate the power continuously after start-up and as long as the fuel supply is provided and the catalysts are active. However, an appropriate thermal integration of the fuel processing components is crucial to minimize the heat losses and ensure high thermal efficiency since we did not consider any heat loss in the process simulation. The combustor has to be integrated with the vaporizer, and the reformer for the thermal coupling. On the other hand, an integration scheme for minimizing the total volume is also desirable. We propose here a hardware integration scheme for the fuel processor including four major components (Vaporizer, Reformer, PrOx and Combustor), several internal heat exchangers and internal thermal barriers. The thermal barriers between the adjacent processing units for bridging the high temperature components from the low ones are required because of the short conduction paths inherent to microreactors. Then, a 2-D CFD simulation for this system was performed to predict the concentration, temperature profiles and the size of each main component.

Chemical kinetics is the key information for the reactor design. There are numerous kinetic studies for the methanol steam reformer [16-19], preferential oxidation of CO [20-22] and hydrogen combustor [23-25]. However, most of them were obtained in macro-scale reactors with limited information about catalyst formulations. In our design, we used literature kinetic data for the steam reformer and the combustor, and in-house kinetics for the PrOx, all of which were then incorporated into CFD package Fluent for simulations. The kinetics for methanol steam reforming is a power-law expression by Samms & Savinell (2002) [16]. In our steam reformer design, we only considered the main reaction (2) and neglected the side reactions (3) & (4). The kinetics for PrOx reactor uses Langmuir-Hinshelwood expressions by Ouyang & Besser [15]. In our PrOx design, we considered three reactions (5~7) in parallel. The kinetics for hydrogen combustor is a one step power-law expression provided by Schefer (1982) [23].

Based on the above kinetics, we estimated the catalyst loading for each reactor considering the process requirement. It turned out that there is an order of magnitude difference on catalyst requirement in each reactor. In order to integrate these reactors at significantly different catalyst loadings, we proposed two types of microreactors with different catalyst synthesis concepts. For the reformer and PrOx, a catalyst cartridge approach microreactor with packed bed or cellular structure catalyst is proposed, whereas for the combustor, a multichannel microreactor with thin-film or a single wire catalyst could be used. A fuel processor hardware integration scheme was then proposed, which includes major process components, heat exchangers and thermal barriers. A 2D CFD simulation was performed to obtain the detailed information about transport phenomena. Figure 2 shows the integration scheme with stream arrangements and heat management. The combustor is sandwiched between the steam reformer and vaporizer since it will provide the heat for both units. The flow in the combustor is co-current with that of the reformer, while counter-current to that of the vaporizer; this arrangement is for the consideration of the temperature distribution and piping. In Figure 2, the arrow between different layers denotes the amount of heat transferred in-between, and the value inside each unit represents the amount of energy available for transfer out of component (negative) or required to transfer to component (positive). The energy values in CFD simulation are consistent with those of process simulation.

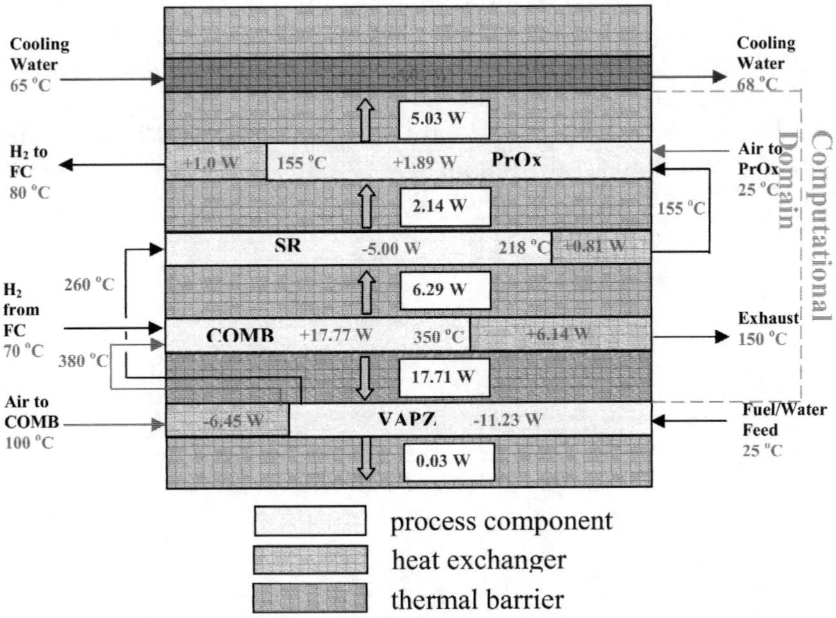

Figure 2 FC integration scheme with stream arrangements and heat management

The size of each processing chip is 60 mm x 70 mm x 0.5 mm. During our CFD simulation, we used 1.5 mm fixed thickness of the thermal barrier and kept adjusting the thermal conductivity (0.01 ~ 0.03 W/mk). The real size will be obtained based on its true thermal conductivity. The simulation results show that the thermal barriers are very important for maintaining the temperature difference, and they occupy a substantial amount of the space. The design of these barriers is critical to the hardware integration scheme.

Figure 3 is the temperature distribution of the simulated system. It includes 7 layers: barrier, combustor, barrier, reformer, barrier, PrOx and barrier (from bottom to top), in which the very bottom surface attaches to the top surface of the vaporizer, and the very top surface contacts the cooling water. The boundary conditions for those two surfaces are the known temperatures, with zero flux imposed on all other solid surfaces. The cooling water can be circulated with the cooling system from the fuel cell stack. From Figure 3, we can see that the temperature inside each reactor is no longer uniform. The axial temperature profiles in these processing components are plotted in Figure 4. The combustor has the most uneven temperature distribution, while the reformer and the PrOx have relatively better temperature profiles. We think that this design is acceptable since the narrow temperature distribution is desired for the steam reformer and the PrOx, whereas this is not critical for the vaporizer. We also notice that there is some difference between the actual stream temperatures and those obtained from the process simulation: the temperature of the air steam to the combustor is preheated to 380 °C from its 100 °C inlet, and the exhaust gas temperature is 150 °C. The preheating of the combustor air is very important because of the high reaction rate near the entrance of the reformer; it needs a fast source of heat. These stream temperature requirements indicate that some external heat exchangers are also required which have not been included in this work.

From CFD simulation, we know that the temperature distribution is altered after the hardware integration. The heat transfer direction between two layers is no longer in one-way direction. In Figure 4, one can see that there are several places of temperature "cross-over" between the processing layers; e.g., between the steam reformer and the combustor, the heat is transferred from the combustor for the first 38 mm length, while the heat transfer is reversed for the remaining 22 mm length. Temperature distributions on cross-sectional cuts in vertical direction are plotted in Figure 5. We can clearly see the function of the thermal barriers — bridging the temperature difference between processing components. From Figure 5, one can also notice the change of heat transfer direction at these locations.

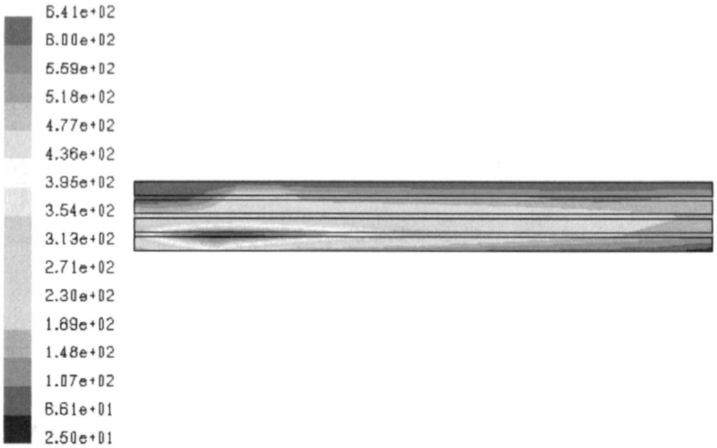

Figure 3 Temperature contour view

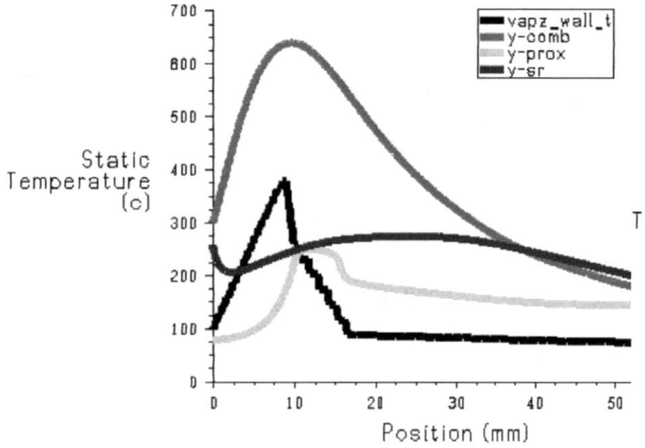

Figure 4 Temperature profile along axial direction in each process layer

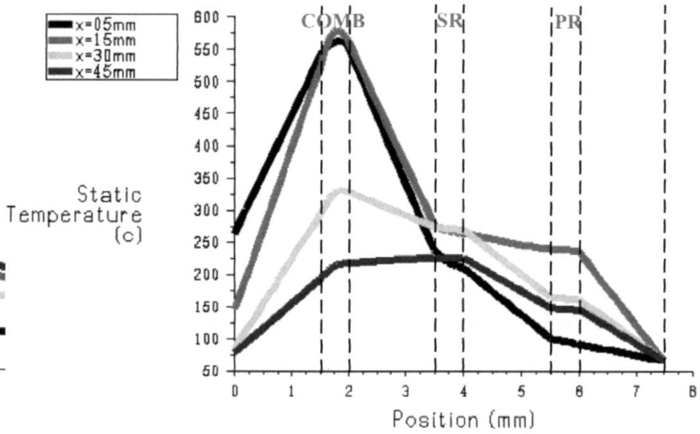

Figure 5 Temperature profiles along 4 vertical lines at different positions

Some detailed information such as wall temperature, heat flux in/out, reaction rate and mass fraction inside the processing components are depicted in Figures 6 ~ 8. From these figures, we can identify the location where the reaction rate is the fastest, and the maximum heat is required or generated. From the mass fraction profiles, we know that this design can meet the process requirements: ~100% conversion of methanol in the reformer, ~ 100% conversion of CO in the PrOx, and ~100% conversion of H_2 in the combustor. The dashed vertical lines specify the position where the reaction is complete, and the rest of the processing component simply serves as a heat exchanger. Therefore, we only need to load the catalyst up to this location.

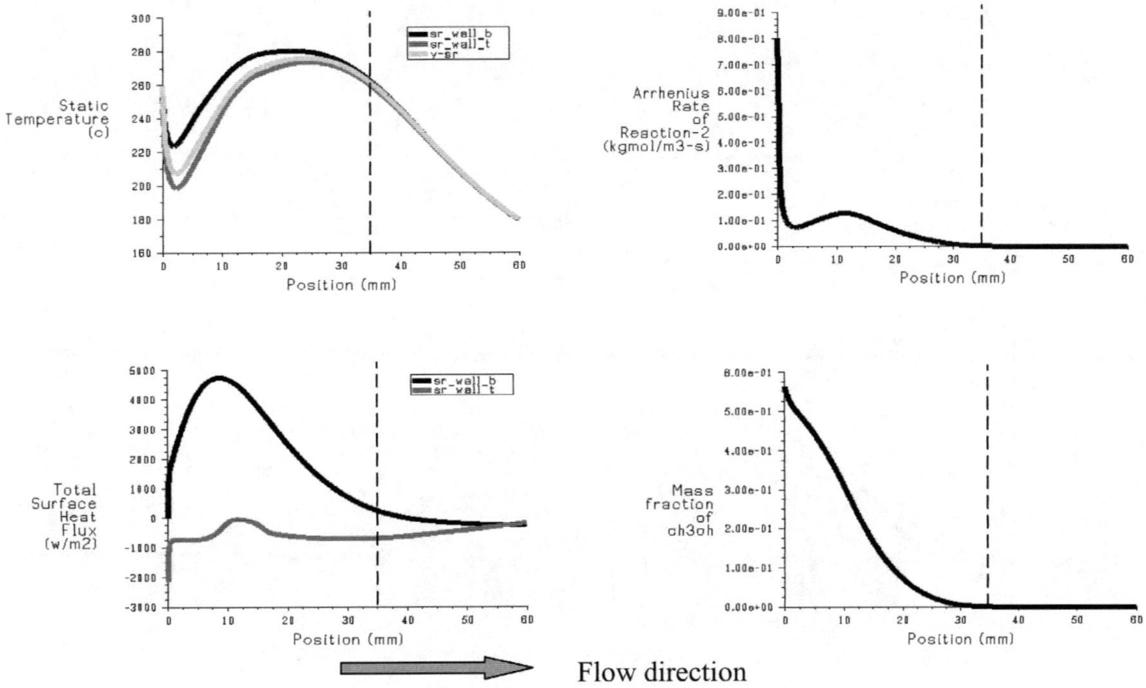

Figure 6 Temperature, heat flux, reaction rate and mass fraction of methanol along the x direction in steam reformer

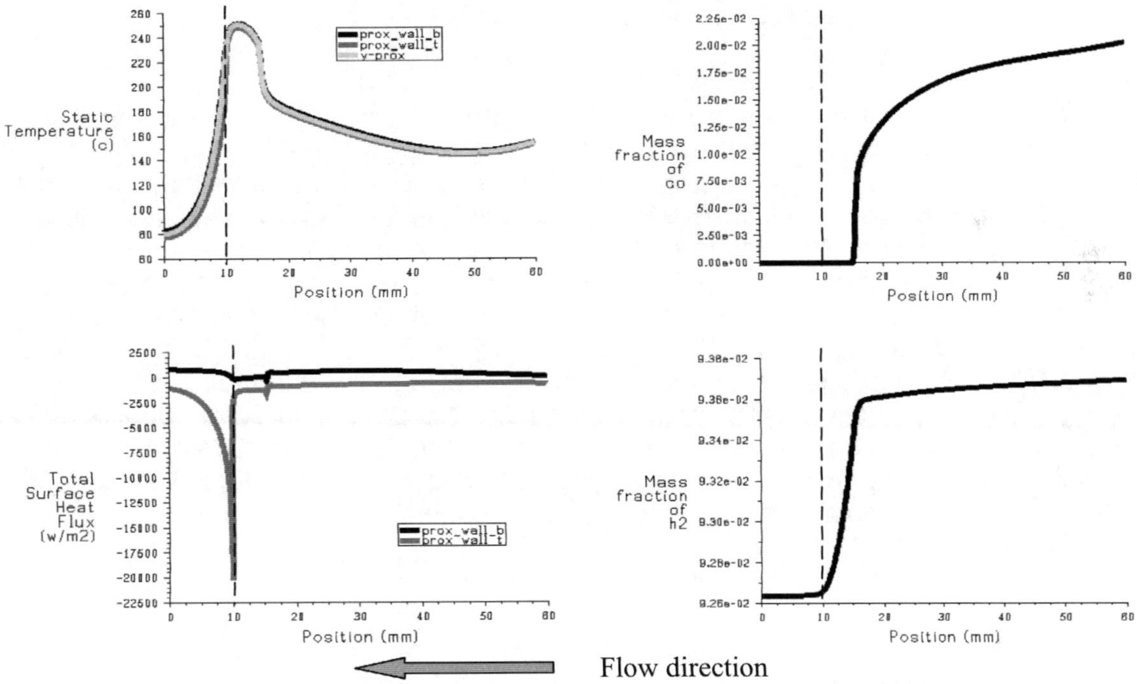

Figure 7 Temperature, heat flux, mass fraction of CO and H_2 along the x direction in PrOx

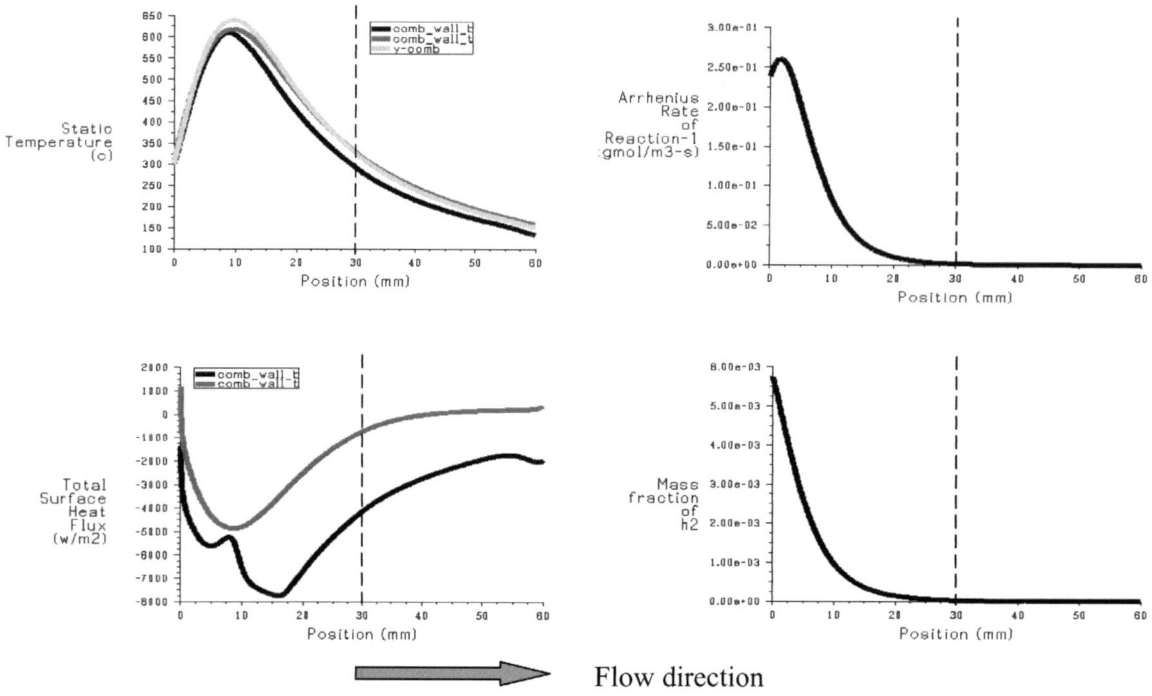

Figure 8 Temperature, heat flux, reaction rate and mass fraction of H_2 along the x direction in combustor

Conclusions

A 20 W PEMFC fuel processor to be incorporated in a fuel cell aimed for military applications based on steam reforming of methanol was simulated and evaluated. Based on process simulation and energy requirement, a microfludic-based hardware integration scheme was proposed and examined. In this PEMFC processing, the steam reforming is the slowest step, so the design of the steam reformer is the key for controlling the total size of the fuel processor. A high performance catalyst for the steam reformer is demanded. The thermal barriers between the processing units are critical for maintaining the temperature difference between the integrated components. The simulation results show that these thermal barriers occupy a significant amount of the total space. A material with low thermal conductivity will help to reduce the total processor volume. Although in the process simulation, an isothermal condition for each reaction unit was assumed, the temperature in each processing unit was not constant after the thermal integration. The state and flow direction of the input streams, the size of the processing components, heat exchangers and thermal barriers were all determined from the CFD simulation.

Acknowledgements

The authors gratefully acknowledge the financial support from Picatinny TACOM/ARDEC under grant No. DAAE30-00-D-1011. Dr. Arthur Kaufman provided a lot of useful suggestions in the discussion of process simulation. We also thank Dr. Woo Lee and Mr. Keyur Shah for the discussion of fuel cell.

Reference

1. Larminie, J. and A. Dicks. (2000), "Fuel Cell Systems Explained", *John Wiley & Sons Ltd,* New York.

2. Park, S., J.M. Vohs, R.J. Gorte (2000), "Direct oxidation of hydrocarbons in a solid-oxide fuel cell", *Nature* 404 (6675), 265–267.

3. Deluga, G.A., J.R. Salge, L.D. Schmidt, X.E. Verykios (2004), "Renewable Hydrogen from Ethanol by Autothermal Reforming", *Science* 303 (5660), 993–997.

4. Appleby, A.J., F.R. Foulkes (1989), Fuel Cell Handbook, *Van Nostrand Reinhold*, New York

5. Thampan, T., S. Malhotra, J. Zhang, R. Datta (2001), "PEM fuel cell as a membrane reactor", *Catal. Today* 67(1-3), 15-32.

6. Loffler, D.G., K. Tylor, D. Mason (2003), "A light hydrocarbon fuel processor producing high-purity hydrogen", *J. Power Sources* 117(1-2), 84-91.

7. Lindstrom, B., L.J. Pettersson (2003), "Development of a methanol fuelled reformer for fuel cell applications", *J. Power Sources* 118(1-2), 71-78.

8. Holladay, J.D., E.O. Jones, R.A. Dagle, G.G. Xia, C. Cao, Y. Wang (2004), "High efficiency and low carbon monoxide micro-scale methanol processors", *J. Power Sources* 131(1-2), 69–72.

9. Cunha, J., J.L.T. Azevedo (2000), "Modelling the integration of a compact plate steam reformer in a fuel cell system", *J. Power Sources* 86(1), 515–522.

10. Delsman, E.R., M.H.J.M. de Croon, G.J. Kramer, P.D. Cobden, Ch. Hofmann, V. Cominos, J.C. Schouten (2004), "Experiments and modelling of an integrated preferential oxidation-heat exchanger microdevice", *Chem. Eng. J.* 101(1-3), 123–131.

11. Ratnamala, G. M., N. Shah, V. Mehta, P.V. Rao, S. Devotta (2005), "Integrated fuel cell processor for a 5-kW proton-exchange membrane fuel cell", *Ind. Eng. Chem. Res.* 44(5), 1535-1541.

12. Godat, J., F. Marechal (2003), "Optimization of a fuel cell system using process integration techniques", *J. Power Sources* 118(1-2), 411–423.

13. Amphlett, J.C., R.F. Mann, B.A. Peppley, D.M. Stokes (1991), "Some design considerations for a catalytic methanol steam reformer for a PEM fuel cell power generating system", *Proceedings of the Intersociety Energy Conversion Engineering Conference* 3, 642-649.

14. Ouyang, X., L. Bednarova, R.S. Besser, P. Ho (2005), "Preferential oxidation (PrOx) in a thin-film catalytic microreactor: Advantages and limitations", *AIChE J.* 51 (6), 1758-1772.

15. Ouyang, X., R.S. Besser (2005), "Effect of reactor heat transfer limitations on CO preferential oxidation", *J. Power Sources* 141 (1) 39-46.

16. Samms, S.R., R.E. Savinell (2002), "Kinetics of methanol-steam reformation in an internal reforming fuel cell", *J. Power Sources* 112 (1), 13-29.

17. Agrell, J., H. Birgersson, M. Boutonnet (2002), "Steam reforming of methanol over a Cu/ZnO/Al2O3 catalyst: A kinetic analysis and strategies for suppression of CO formation", *J. Power Sources* 106 (1-2), 249-257.

18. Peppley, A., J.C. Amphlett, L.M. Kearns, R.F. Mann (1999), "Methanol-steam reforming on Cu/ZnO/Al2O3 catalysts. Part 2. A comprehensive kinetic model", *Applied Catalysis A: General* 179 (1-2), 31-49.

19. Jiang, C.J., D.L. Trimm, M.S. Wainwright, N.W. Cant (1993), "Kinetic mechanism for the reaction between methanol and water over a Cu-ZnO-Al2O3 catalyst", *Applied Catalysis A: General* 97(2), 145-158.

20. Voltz, S.E., C.R. Morgan, D. Liederman, S.M. Jacob (1973), "Kinetic study of carbon monoxide and propylene oxidation on platinum catalysts", *Ind. Eng. Chem. Prod. Res. Dev.* 12(4), 294-301.

21. Kahlich, M.J., H.A. Gasteiger, R.J. Behm(1997), " Kinetics of the selective CO oxidation in H2-rich gas on Pt/Al2O3", *J. Catalysis* 171(1), 93-105.

22. Kim, D.H., M.S. Lim (2002), "Kinetics of selective CO oxidation in hydrogen-rich mixtures on Pt/alumina catalysts", *Applied Catalysis A: General* 224(1-2), 27-38.

23. Schefer, R.W. (1982), "Catalyzed combustion of H2/air mixtures in a flat plate boundary layer - 2. Numerical model", *Combustion and Flame* 45(2), 171-190.

24. Ikeda, H., P.A. Libby, F.A. Williams (1993), "Catalytic combustion of hydrogen - Air mixtures in stagnation flows", *Combustion and Flame* 93(1-2), 138-148.

25. C. Appel, J. Mantzaras, R. Schaeren, R. Bombach, A. Inauen, B. Kaeppeli, B. Hemmerling, A. Stampanoni (2002), "An experimental and numerical investigation of homogeneous ignition in catalytically stabilized combustion of hydrogen/air mixtures over platinum", *Combustion and Flame* 128(4), 340-368.

96d Air-Breathing Laminar Flow Fuel Cells Operating in Alkaline or Acidic Media

Paul J.A. Kenis[1], Ranga S. Jayashree[1], Daniela Egas[1], Dilip Natarajan[2], Jacob Spendelow[1], and Larry J. Markoski[2]. (1) Chemical & Biomolecular Engineering, University of Illinois at Urbana-Champaign, 600 S. Mathews Ave, Urbana, IL 61801, (2) INI Power Systems, 136 Quade Drive, Cary, NC 27513

Laminar flow-based fuel cells (LFFCs) utilize the occurrence of laminar flow at the microscale to create a membraneless fuel cell [1]. Mixing is governed by slow molecular diffusion only, so streams of fuel and oxidant can be kept separated without use of a polymer electrolyte membrane (e.g. Nafion). Compared to more conventional fuel cells (e.g. DMFCs), these LFFCs exhibit the following advantages: (i) fuel crossover can be avoided completely leading to high open cell potentials; (ii) anode dry-out and cathode flooding issues are eliminated; (iii) these cells can be operated in both acidic and alkaline conditions [2]; and (iv) one can operate these fuel cells with either formic acid or methanol as the fuel. Earlier generations of these LFFCs were performance limited by slow mass transport of dissolved oxygen to the cathode, but recently we have overcome this issue by incorporating an air-breathing gas-diffusion electrode which enables rapid transport oxygen to the cathode from air due to the 4 orders of magnitude higher diffusion constant [3].

This presentation will summarize the key characteristics of this microfluidic fuel cell. Then the performance of direct methanol LFFCs with gas diffusion electrodes in acidic and alkaline media will be compared. These cells exhibit high open circuit potentials of 0.93 V and 1.05 V in acidic and alkaline media, respectively, which demonstrates control over methanol crossover in these fuel cell designs. A higher maximum power density of 17.2 mW/cm2 was obtained when operating in alkaline media compared to 11.8 mW/cm2 in acidic media. Single electrode studies show that the cathode and anode exhibits better kinetics in alkaline solutions, in accordance with prior electrochemical studies in literature. These LFFCs, however, avoid clogging of membrane pores by carbonate formation in alkaline media, one of the factors that has limited their use in, for example, DMFCs.

[1] E.R. Choban, L.J. Markoski, A. Wieckowski, and P.J.A. Kenis, J. Power Sources 2004, 128, 54; R. Ferrigno, A.D. Stroock, T.D. Clark, M. Mayer, and G. M. Whitesides, J. Am. Chem. Soc., 2002, 124, 12930. [2] E.R. Choban, J.S. Spendelow, L. Gancs, A. Wieckowski, and P.J.A. Kenis, Electrochim. Acta, 2005, 50, 5390; J. L. Cohen, D.J. Volpe, D.A. Westly, A. Pechenik, and H.D. Abruna, Langmuir, 2005, 21, 3544. [3] R.S. Jayashree, L. Gancs, E.R. Choban, A. Primak, D. Natarajan, L.J. Markoski, and P.J.A. Kenis, J. Am. Chem. Soc., 2005, In press

Addressing Critical Issues in Microscale Fuel Processing: Demonstration of An Integrated Silicon Microreactor Based Methanol Steam Reformer

Keyur Shah and R.S. Besser
New Jersey Center for Microchemical Systems, Chemical, Biomedical and Materials Engineering Department, Stevens Institute of Technology, Hoboken, NJ 07030
Correspondence: Prof. R.S. Besser, (201) 216-5257 (voice), (201) 216-8306 (fax), rbesser@stevens.edu

As fuel cells gain attention as a potential energy solution of the future, one of the most promising fields of application is portable power, as a prominent alternative to batteries. In applications where the power requirements are low but weight and volume are critical, it is believed that miniature fuel cells will be preferred over conventional batteries [1-3]. However, the success of fuel cell technology for portable power depends heavily on development of an efficient way of delivering fuel to the cell. Methanol reforming offers an attractive source of hydrogen for fuel cell systems for portable power. The greatest obstacle to realization of these systems is the extraction of hydrogen from the hydrocarbon fuel, i.e., the fuel processing part. Considerable efforts have been made to develop an integrated micro fuel processor/fuel cell system, however, development of a miniature chemical system that can effectively exploit the high energy density of methanol and other liquid hydrocarbons offers a number of significant technological challenges. Miniaturization of system components and thermal management in miniature systems are perhaps the most crucial challenges for microscale fuel processors. Through this work, we are trying to address these crucial system issues.

Methanol (CH_3OH) is an attractive fuel for portable power, offering high hydrogen-carbon ratio, high energy density (5600 Wh/kg), and ready availability [4]. Moreover, methanol is sulfur free and can be reformed relatively easily at low temperatures (250-300°C), simplifying the micro fuel processor design. Figure 1 shows a schematic for processing of methanol to produce hydrogen-rich gas suitable for a low temperature (70-80°C) proton exchange membrane (PEM) fuel cell.

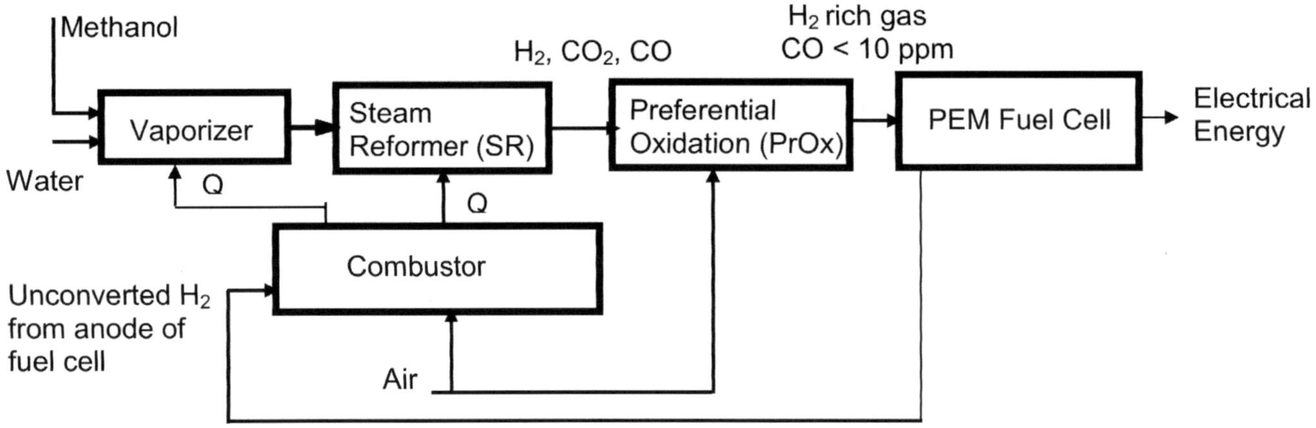

Figure 1. Processing steps required for generating high purity hydrogen from methanol

Because of their size and portability, microchemical systems have shown promising results in chemical process miniaturization. Microreactors are becoming a strong enabling aspect of compact fuel processors due to their advantageous heat and mass transport and inherent compactness [5-8]. Microreaction can facilitate highly compact power sources through integration of a fuel cell (FC) with all the unit operation components of a fuel processor (FP) along with microstructured sensors, actuators and other "balance of plant" (BOP) devices.

We are exploring silicon microfabrication technology to achieve miniaturization. Silicon is lighter than stainless steel; it possesses high mechanical strength, chemical inertness, thermal stability, and large thermal conductivity. With the established silicon microfabrication technology, it is possible to create novel microchannel and reactor configurations in the sub-millimeter range with tight dimensional controls. An important advantage of silicon microfabrication is its compatibility with thin film technology. This facilitates eventual on-chip integration of functional elements like sensors and actuators allowing compact process control subsystems. However, it is our intent at this stage to perform experiments whose outcome depends most critically on microscale geometry and less so on the materials of construction. In this way the results obtained may be generalizable to any of the materials technologies.

Thermal Integration Issues in Micro Fuel Processor

The management of heat in a compact format is perhaps the most crucial challenge for micro scale fuel processors. The fuel processor needs effective thermal coupling to allow transfer of energy from the heat producing combustor to the endothermic steam reformer (SR). Coupling endothermic and exothermic components of the fuel processor and minimizing losses can achieve a high thermal efficiency. However, such coupling must be accomplished in a manner that permits the maintenance of specific temperatures in the various components as shown in Figure 2 and maintains the surface of the package near room temperature.

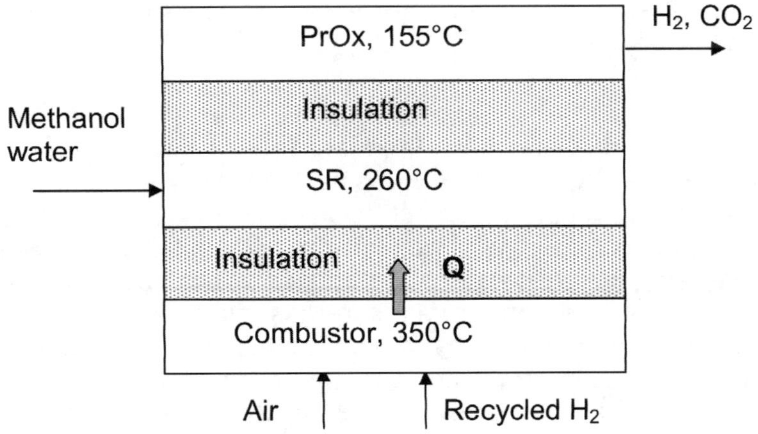

Figure 2. Integration of fuel processing components

In this study, a silicon microreactor-based catalytic methanol SR was designed and fabricated in the context of complete thermal integration to directly address the heat management issue. As shown in Figure 3, a silicon micro SR is designed and integrated with microfabricated insulation as dictated by overall system design. A thin-film microfabricated heater is used to simulate integration with a combustor by controlled heat introduction.

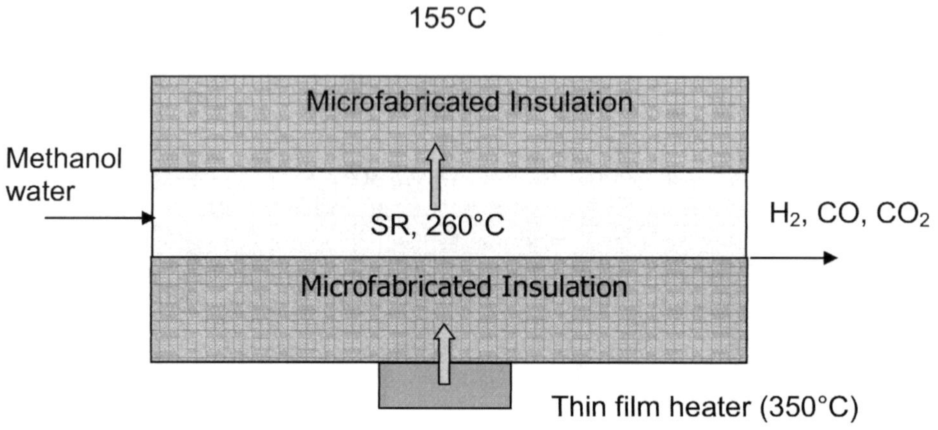

Figure 3. Demonstration of thermally integrated SR microreactor

Thermal Coupling and Insulation

Because of the short conduction paths inherent to microreactors, high performance insulators are needed between fuel processing components to isolate the high temperature components. The objective is to have insulating materials that can offer ultra low thermal conductivity to bridge these temperature differences and to minimize conductive and convective heat losses to the surroundings without adding much to the weight and volume of the total system.

The best commercial insulators have thermal conductivities in the range of 0.02 W/mK and are difficult to integrate into a microfabricated unit. However, vacuum packaging of microreactors can provide an effective means of insulation. From the kinetic theory of gases, it is known that the thermal conductivity of a gas is approximately independent of pressure for atmosphere and above, and decreases with sub-atmospheric pressures [9]. A microfabricated cavity surrounding the reactor filled with low-pressure gas offers low thermal conductivity down to approximately 0.001W/mK and can be engineered by managing pressure and cavity depth. Silicon microfabrication enables a straightforward approach to this structure as a cavity of precise depth can be etched in silicon and bonded with either silicon or glass using vacuum bonding methods [10]. Figure 4 shows a concept for microfabricated vacuum insulation.

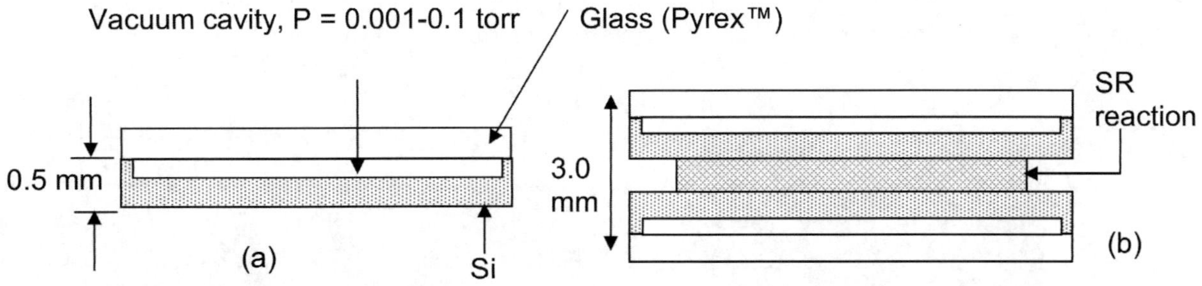

Figure 4. (a) Vacuum insulation chip with a single microfabricated cavity in Si, and (b) SR reaction zone sandwiched between insulation chips

Integrated Micro SR: Design and Fabrication

The design of an integrated SR is shown in Figure 5, where vacuum packaging chips, thin film metal heaters and temperature sensors are directly embedded with the reforming microreactor to simulate an integrated SR in an overall fuel processing scheme.

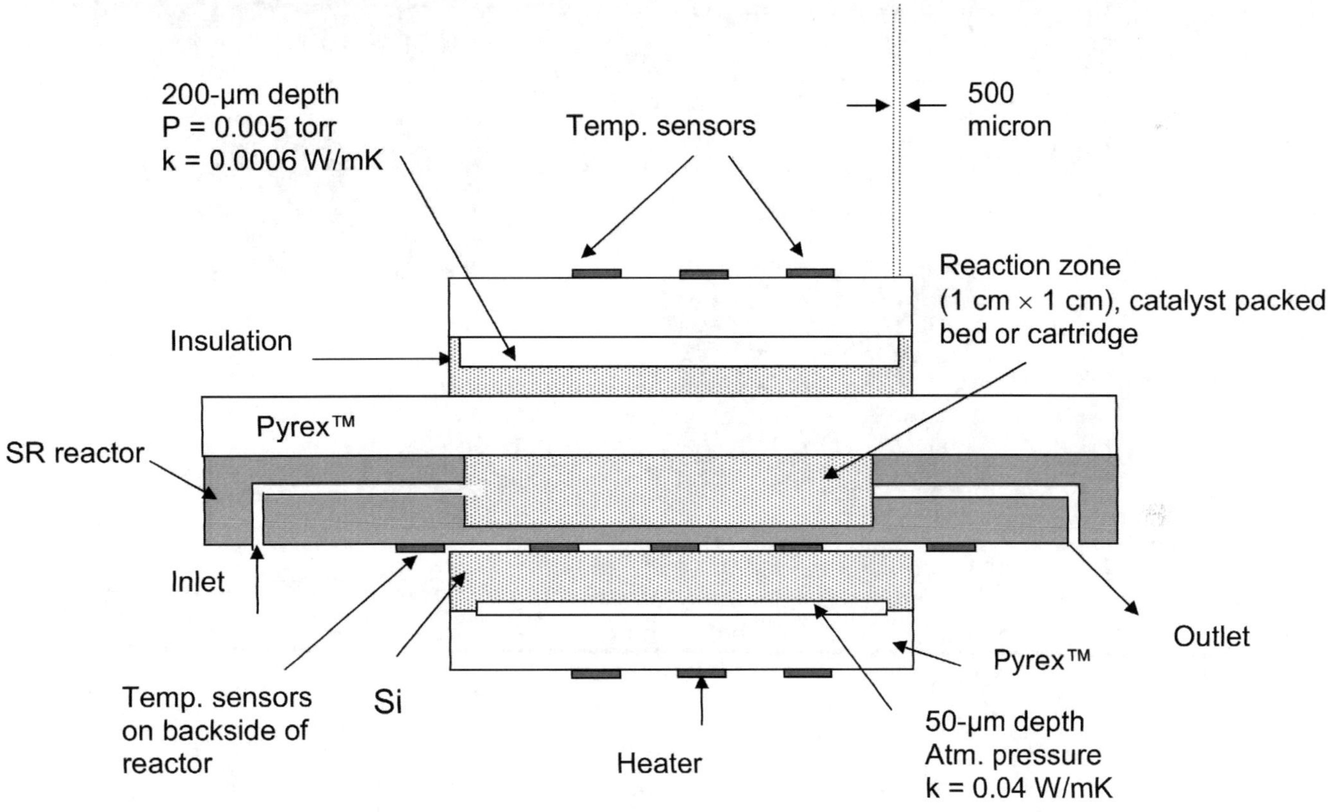

Figure 5. Design of an integrated SR

Silicon microfabrication technology is used to fabricate reforming microreactors, vacuum packaging chips, thin-film heaters and temperature sensors on silicon and glass substrates. Figure 6 shows fabricated components of this integrated device.

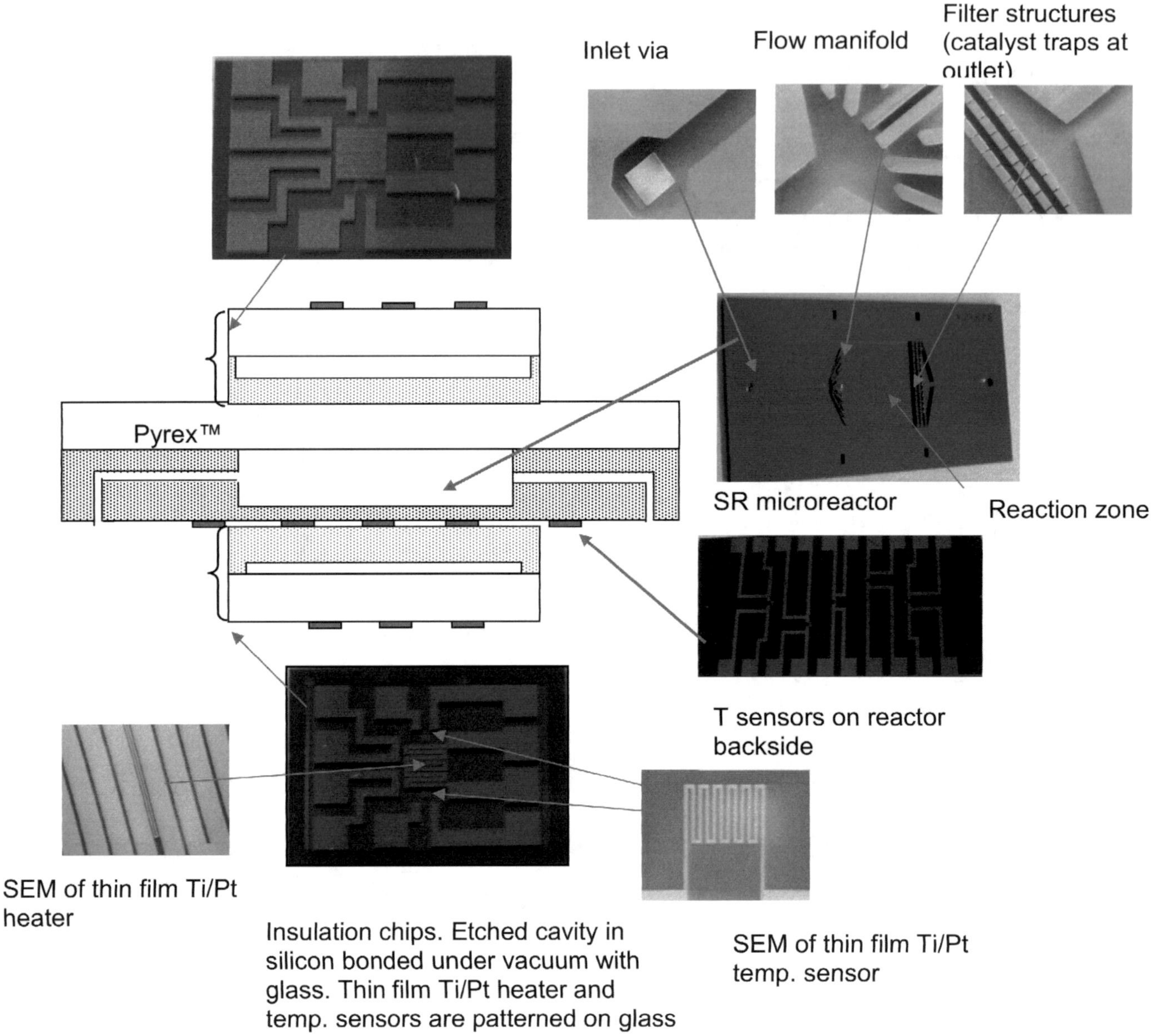

Figure 6. Fabrication of an integrated SR using silicon microfabrication techniques

The catalyst is incorporated in the form of a packed bed in the silicon microreactor chip bonded with glass. Figure 7 shows the silicon microreactor packed with catalyst particles of mean diameter 70-80 μm.

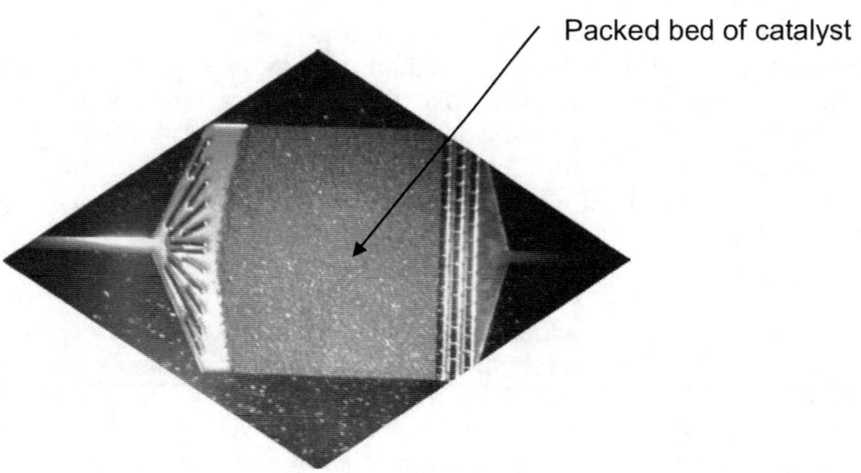

Figure 7. Incorporation of catalyst in the form of packed bed by vacuum loading
Catalyst loading achieved: 51 mg

Next, the insulation chips are bonded on the front and back sides of the catalyst-loaded microreactor using high temperature epoxy. Figure 8 shows the microreactor bonded with insulation chips.

Frontside of SR microreactor bonded with insulation chips

Backside of SR microreactor bonded with insulation chips

Assembled integrated microreformer (six-chip stack)

Figure 8

Thermal Characterization

An important aspect of integration was the thermal management at the system level, which involved heat required, transfers of heat between units, temperature profiles in different components, insulation effectiveness, and heat loss mechanisms. Prior to detailed design of individual reactors, necessary data were needed to confirm the accuracy of the assumptions used to simulate the thermal behavior. Therefore fabricated model components of an integrated device were used for the acquisition of critical thermal parameters of the overall coefficients for the transfer of heat from system fluids to one another and to the outside environment by different means of heat loss.

Thermal characterization experiments were carried out under different conditions at both atmospheric pressure and under vacuum to distinguish multiple modes of heat transport. These experiments helped quantify heat losses through various pathways from the planar microreactor structure. Well defined external heat flux is introduced to simulate the flows that result from integration with the combustor. Distributed sensors are provided to obtain the temperature profiles along the axial length.

The following sets of experiments were carried out.

1. Device suspended under ambient pressure
2. Device suspended in vacuum chamber
3. Device covered with commercial insulator under ambient pressure
4. Device covered with commercial insulator in vacuum
5. Device covered with Al foil and open to ambient
6. Device covered with Al foil in vacuum chamber

For each experiment, $Q_{combustor}$ required to raise the temperature of SR to 260°C was measured. Figure 9 and Figure 10 show thermal characterization results when tested under atmospheric pressure and vacuum respectively.

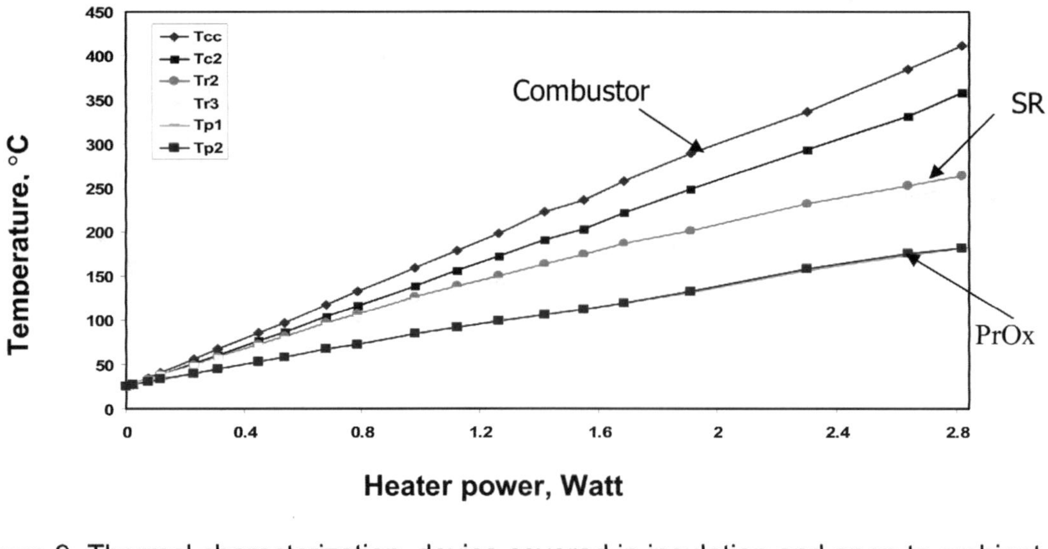

Figure 9. Thermal characterization, device covered in insulation and open to ambient. Heat required versus temperature profiles in SR, combustor, and PrOx

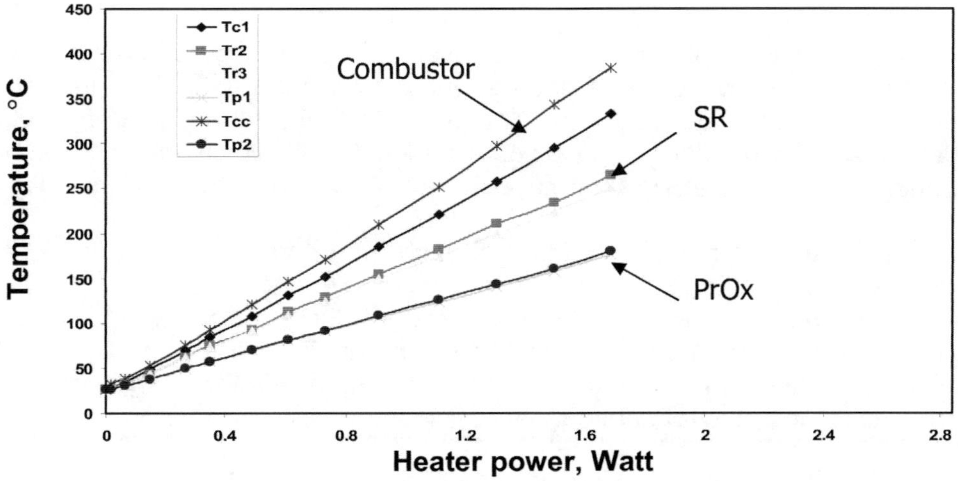

Figure 10. Thermal characterization, device covered in insulation and placed in vacuum chamber. Heat required versus temperature profiles in SR, combustor, and PrOx

Reaction Characterization

Experiments are carried out to determine conversion and selectivity, hydrogen production rate (corresponding power output), heat required to drive the reforming reaction, and thermal effects on reaction. Steam reforming reaction was carried out at different total flow of methanol+water (H_2O: MeOH=1.3) from 0.007 ml/min - 0.015 ml/min at 260°C. Figure 11 shows methanol conversion obtained at different flow rates of combined feed. Since the heat required to carry out endothermic SR reaction is quite small relative to amount of heat to be input (because of high heat losses) in order to achieve required reaction temperature in SR, we did not observe any major change in temperature profiles of SR microreactor, combustor and PrOx during the reaction.

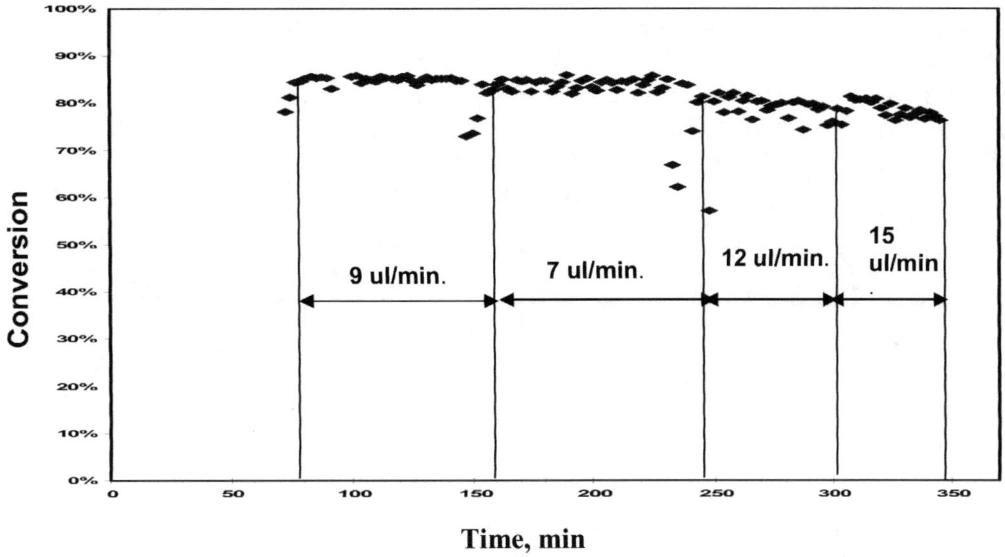

Figure 11. MeOH conversion obtained at different flow rates of water, MeOH feed

Based on the results obtained from above thermal characterization experiments, modeling is under progress to quantify heat loss mechanism and to deduce free convection heat coefficients from microscale geometry. These data will then help to verify and improve thermal model. Based on understanding gained, suggestions will be made for scale up of reactor components and a packaging scheme for reduction of convective and radiative losses.

This combined experimental/modeling approach then can be used as the quantitative basis for designing an integrated portable fuel processor. The refined model can be used for energy density projections of portable fuel processor/fuel cell systems as well as for projecting energy requirements to microchemical reactors in general. The research outcome will help conceive applications and limitations of planar microchemical technology for micro fuel processing and provide impetus for subsequent development.

Acknowledgements

The authors acknowledge the financial support from the Picatinny TACOM/ARDEC and New Jersey Commission on Science and Technology (NJCST).

The silicon and glass microfabrication was performed in part at the Cornell NanoScale Facility (a member of the National Nanofabrication Users Network) which is supported by the National Science Foundation under Grant ECS-9731293, Cornell University and industrial affiliates.

References

1. C.K. Dyer, *Journal of Power Sources* **2002**, 106, 31.
2. Keyur Shah, W.C. Shin, R.S. Besser, *Journal of Power Sources* **2003**, 123, 172.
3. Keyur Shah, W.C. Shin, R.S. Besser, *Sensors & Actuators B* **2004**, 97, 157.
4. Y. Choi, H.G. Stenger, *Applied Catalysis; B* **2002**, 38, 259.
5. J.D. Holladay, E.O. Jones, M. Phelps, J. Hu, *J. Power Sources* **2002**, *108*, 21.
6. A.V. Pattekar, M.V. Kothare, *Journal of Microelectromechanical system* **2004**, *13(1)*, 7.
7. R.S. Besser, X. Oyuang, H. Surangalikar, *Chemical Engineering Science* **2003**, 58, 19.
8. H. Surangalikar, X. Ouyang, R.S. Besser, *Chemical Engineering Journal* **2003**, 93, 217.
9. P. Atkins, J.D. Paula, *Physical Chemistry*, 7th ed., W.H. Freeman & Company, NY **1997**.
10. Keyur Shah, X. Ouyang, and Ronald S. Besser, *Chemical Engineering Technology,* **2005**, 28(3), 303.

132a Long-Term Sulphation Behavior of Limestone Sorbents

Jinsheng Wang and Edward J. Anthony. CANMET Energy Technology Centre, 1 Haanel Dr., Ottawa, ON K1A 1M1, Canada

Limestone particles are used in fluidized bed combustion (FBC) as sulphur sorbents. The rate of sulphation of the sorbents decreases quickly with sulphation time, and it is well observed that under typical conditions, the limestones will attain an effective maximum sulphation conversion in several hours and cease to significantly reacts further. The concept of a maximum conversion level has been employed in many modeling efforts. A theoretical maximum conversion level which is below 100 % has also been proposed. However, recent observations indicate that there is no such upper limit and 100 % conversion can be achieved over long time of sulphation. In industrial FBC boilers large sorbent particles can have a residence time up to 20 hours. It has been suggested that long-term sulphation of the large particles could make a significant contribution to the sulphur capture. Understanding the behavior of long-term sulphation is also of importance for operation and maintenance of FBC boilers. Long-term sulphation could cause agglomeration of the sorbent particles and formation of hard deposits in FBC boilers firing high sulphur coals or petroleum cokes. As a result, severe fouling could develop in the boilers. While short-term sulphation of limestones have been extensively studied, there are only limited studies on the long-term sulphation. The behavior of long-term sulphation is quite different from that of the short-term one, and it is believed that short-term sulphation conversion seen in a typical thermogravimetrical analysis (TGA) tests has no relationship to the long-term conversion. Based on sulphation tests of a few limestones, equations have been proposed to describe the time evolution of the sulphation conversion. However, these equations could not cover the early or rapid sulphation stage. Accordingly, for description of the total sulphation period at least two separate equations are required. Moreover, the starting point of the long-term stage is not well defined. In this work we propose a single equation for the sulphation behavior for any period of time. At short sulphation periods, the equation reduces to a well known form. For long sulphation periods, the equation predicts complete conversion and thus conforms to the observation. The equation has been applied to various long-term sulphation data and a satisfactory description is achieved.

132b Novel Sorbents for Mercury Capture from Fuel Gas

Evan J. Granite, Christina Myers, Dennis Stanko, and Henry W. Pennline. United States Department of Energy, National Energy Technology Laboratory, PO Box 10940, M/S 58-106, Pittsburgh, PA 15236-0940

There are increasing concerns over mercury pollution in the environment. Mercury is a powerful neurotoxin, and can bio-accumulate up the food chain. The United States Environmental Protection Agency (USEPA) announced on December 14, 2000 its intention to regulate coal-utilizing (gasifiers and combustors) facilities for mercury emissions. The USEPA issued the regulation in March 2005. With EPA's Clean Air Mercury Rule, and several states promulgating their own regulations, the need exists for a low cost mercury removal technique that can be applied to integrated gasification combined cycle (IGCC) and coal-burning power plants. Gasification is an important strategy for increasing the utilization of abundant domestic coal reserves. The Department of Energy envisions increased use of gasification in the United States during the next twenty years. As such, the DOE Gasification Technologies Program will strive to approach a near-zero emissions goal with respect to pollutants such as mercury. In gasification for power generation, the removal of mercury by sorbents at elevated temperatures preserves the high thermal efficiency of the integrated gasification combined cycle system. Unfortunately, most sorbents will display poor capacity for elemental mercury at elevated temperatures. Previous experience with sorbents in flue gas has allowed for judicious selection of potential high temperature candidate sorbents. The capacities for many sorbents for elemental mercury from nitrogen, as well as from four different simulated fuel gases at temperatures from 400F - 700F, were determined. The simulated fuel gas compositions contain varying concentrations of carbon monoxide, hydrogen, carbon dioxide, moisture, and hydrogen sulfide. Promising high temperature sorbents have been identified. Future directions for mercury sorbent development for fuel gas application will be discussed.

132c The Pco Process for Photochemical Removal of Mercury from Flue Gas

Evan J. Granite[1], Henry W. Pennline[1], and Christopher R. McLarnon[2]. (1) United States Department of Energy, National Energy Technology Laboratory, PO Box 10940, M/S 58-106, Pittsburgh, PA 15236-0940, (2) Powerspan Corp., PO Box 219, 54 Old Bay Road, New Durham, NH 03855

The National Energy Technology Laboratory has developed several promising techniques to capture and remove elemental mercury from flue gases. Powerspan Corp. has licensed one of these technologies for application to coal-derived flue gases, and has initiated a bench and pilot test program to develop the Photochemical Oxidation, or PCO Process for commercial application with subbituminous and lignite fuels.

The PCO Process has the potential to serve as a low cost mercury oxidation technology that will facilitate the removal of elemental mercury within a downstream SO2 scrubber, wet electrostatic precipitator (WESP), or baghouse. The technique employs 254-nanometer (nm) ultraviolet light to produce an excited state mercury species in the flue gas, leading to oxidation of elemental mercury. Testing conducted in Powerspan's laboratories showed greater than 90% oxidation and removal of elemental mercury from a realistic simulated subbituminous flue gas stream at bench-scale. The parasitic power requirements are small and estimated to be less than 0.5%. Pilot-scale testing of the PCO Process is planned to begin soon.

132d Performance Optimization and Emission Reduction of a Baled Biomass Combustor

Michael E. Harasek, Martin Miltner, Angela Miltner, Christian Jordan, and Anton Friedl. Institute of Chemical Engineering, Vienna University of Technology, Getreidemarkt 9/166, Vienna, AK A-1060, Austria

The use of biomass for combustion processes is of increasing interest for the production of heat and power. The development of appropriate designs of such combustion systems for these fuels is rather challenging, often time-consuming, and cost intensive due to the heterogeneous composition and disadvantageous ash melting behaviour of biomass. The application of computational fluid dynamics (CFD) offers the possibility to reduce development costs and the time-to-market of the combustion process and its apparatuses and machinery. The development of an innovative system based on an adiabatic combustion chamber with three distinct zones - primary firing zone, post combustion zone, and grate furnace - will be presented. A new model for the formation of volatile and NOx precursors within the biomass bale has been implemented into the CFD package FLUENT. Simulation result will be compared with experiments

132e Modeling and Simulation of Diesel Particulate Filters

Herbert DaCosta, Advanced Materials Technology/T&SD, Caterpillar Inc., Technical Center-E/854, P.O. Box 1875, Peoria, IL 61656

On-highway vehicles powered with diesel engines will need to comply with increasingly stringent EPA regulations for particulate matter and NOx emissions in 2007 and 2010. In order to reach the required emission standards, modeling and simulation approaches can provide significant help in designing new systems and in optimizing operating conditions. Diesel particulate filter (DPF) is the technology of choice to control particulate emissions. In this work, we use various simulation methods to model particle deposition and regeneration in DPFs.

146a Desalination and Water Reuse Technologies in the Petroleum Refining Industry

Tom Pankratz, CH2M Hill, P.O. Box 75064, Houston, TX 77234

Water is essential to many petroleum refinery operations. In fact, a typical refinery handles more water than oil for a variety of heating, cooling and process operations. As fresh water resources face growing pressures, alternative supply sources have to be considered and new, more effective treatment technologies must be employed to meet increasingly strict quality limitations placed on industrial wastewater discharges. Fortunately, advances in treatment technologies – particularly membrane and desalination technologies – are providing new treatment options.

Maturation of microfiltration (MF), ultrafiltration (UF), nanofiltration (NF) and reverse osmosis (RO) have proven effective at reducing water treatment costs while simultaneously improving treated water quality. The application of an appropriate membrane process often makes it practical to treat marginal and brackish water sources that would not otherwise have been considered. Even seawater desalination is proving a viable option at many refineries around the world.

Membrane technologies have also increased wastewater treatment options. The membrane bioreactor (MBR) process – which combines an aerobic biological process with an integrated, low-pressure membrane system – is an accepted wastewater treatment technology that is now being considered for many new wastewater treatment plants and plant upgrades. MBRs produce a higher quality effluent, reduce plant footprints, and provide effective pretreatment if a subsequent RO process is to be considered.

These new technologies have led to a dramatic increase in the use of reclaimed wastewaters within petroleum refineries. Dual membrane systems integrating low-pressure MF/UF with RO may be used to repurify secondary or tertiary effluents to virtually any desired level of quality; water suitable for boiler make-up and/or a variety of process water applications. In addition to treating their own internally produced wastewaters, many refineries are turning to repurified municipal wastewater as a 'new' water source.

Membrane processes may be combined with thermal treatment systems to achieve zero or near zero liquid discharge. These hybrid arrangements often result in reduced energy consumption and in some cases, are able to recover waste heat as an energy source.

An increasing number industrial water supply projects now consider desalination and water reuse as economically competitive, drought-proof water management options. This presentation will review the membrane and evaporative technologies that are making this possible, and will consider how they may be being applied within a petroleum refinery.

The information will be supported by the presentation of case histories and operational data from refineries in the USA, Australia, the Caribbean, India, and Saudi Arabia.

146b Waste Heat and Thermal Desalination

John Tonner, Water Consultants International, 10201 North Concord Drive, Mequon, WI 53097

The largest application of thermal desalination is the use of a desalination plant as the main condenser for a steam power plant. This technique using steam from condensing or back-pressure turbines is widely used in the middle east where single distillers can produce up to 20 million gallons per day (MGD) of fresh water and dozens of such machines may comprise a single facility. The energy consumption to drive a distillation process is high when considered as a stand alone process, but dual purpose generation of power and water yields higher thermodynamic efficiency than making water and electricity separately.

Smaller versions of this middle eastern concept exist in other locations such as the Caribbean where higher energy values mean those systems utilize more efficient, higher capital cost configurations to reach an economic optimum water cost.

Cogeneration of electric power and water may be the most common method of using waste heat for desalination but it is far from the only combination employed. Many merchant marine and cruise vessels produce warm water from engine cooling systems and that warm water is used in distillers to make the ship's water supply. In some cases steam and warm water is used. Naval vessels, particularly nuclear carriers, employ similar technique where much of the water is used for on-deck steam catapults.

There are also industrial applications which may utilize several different types of distiller and feed water, to produce water which can be used by the industry itself, or connected to the local water supply network. In these cases the efficiency of the distiller is usually customized to match the demand for water with the heat available. Newer applications of this type utilize not only traditional heat from stacks and exhausts but also exothermic heat of reactions or are integrated more closely with the industrial processes than ever before.

Oil refineries as well as LNG, Methanol, sulphuric and pteripthalic acid production facilities, have all been coupled with various thermal desalination processes to ensure the water needs of the facility (and its processes) are met. Some examples of these configurations are explained in the text..

The high quality water produced by distillers is usually well matched to the quality required for most industrial processes - including steam generation. Producing this quality of water from the sea, in a single process step, while utilizing otherwise neglected heat sources, is not only environmentally friendly but economically competitive with the ever increasing costs and constantly dwindling volume of traditional water supplies.

American Institute of Chemical Engineers
3 Park Avenue
New York, New York 10016-5991

ISBN: 978-1-60423-529-6

Sensors

Topical Conference at the 2010 AIChE Annual Meeting

Salt Lake City, Utah, USA
7-12 November 2010

ISBN: 978-1-61782-140-0

AIChE